Bioindicator Systems for Soil Pollution

NATO ASI Series

Advanced Science Institutes Series

A Series presenting the results of activities sponsored by the NATO Science Committee, which aims at the dissemination of advanced scientific and technological knowledge, with a view to strengthening links between scientific communities.

The Series is published by an international board of publishers in conjunction with the NATO Scientific Affairs Division

A Life Sciences **B Physics**	Plenum Publishing Corporation London and New York
C Mathematical and Physical Sciences **D Behavioural and Social Sciences** **E Applied Sciences**	Kluwer Academic Publishers Dordrecht, Boston and London
F Computer and Systems Sciences **G Ecological Sciences** **H Cell Biology** **I Global Environmental Change**	Springer-Verlag Berlin, Heidelberg, New York, London, Paris and Tokyo

PARTNERSHIP SUB-SERIES

1. Disarmament Technologies	Kluwer Academic Publishers
2. Environment	Springer-Verlag / Kluwer Academic Publishers
3. High Technology	Kluwer Academic Publishers
4. Science and Technology Policy	Kluwer Academic Publishers
5. Computer Networking	Kluwer Academic Publishers

The Partnership Sub-Series incorporates activities undertaken in collaboration with NATO's Cooperation Partners, the countries of the CIS and Central and Eastern Europe, in Priority Areas of concern to those countries.

NATO-PCO-DATA BASE

The electronic index to the NATO ASI Series provides full bibliographical references (with keywords and/or abstracts) to more than 50000 contributions from international scientists published in all sections of the NATO ASI Series.
Access to the NATO-PCO-DATA BASE is possible in two ways:

– via online FILE 128 (NATO-PCO-DATA BASE) hosted by ESRIN, Via Galileo Galilei, I-00044 Frascati, Italy.

– via CD-ROM "NATO-PCO-DATA BASE" with user-friendly retrieval software in English, French and German (© WTV GmbH and DATAWARE Technologies Inc. 1989).

The CD-ROM can be ordered through any member of the Board of Publishers or through NATO-PCO, Overijse, Belgium.

Series 2: Environment – Vol. 16

Bioindicator Systems for Soil Pollution

edited by

Nico M. van Straalen
Department of Ecology and Ecotoxicology,
Vrije Universiteit,
Amsterdam, The Netherlands

and

Dmitri A. Krivolutsky
Institute of Ecology and Evolution,
Russian Academy of Sciences,
Moscow, Russia

Kluwer Academic Publishers

Dordrecht / Boston / London

Published in cooperation with NATO Scientific Affairs Division

Proceedings of the NATO Advanced Research Workshop on
New Approaches to the Development of Bioindicator Systems for Soil Pollution
Moscow, Russia
24–28 April 1995

A C.I.P. Catalogue record for this book is available from the Library of Congress

ISBN 0-7923-4175-9

Published by Kluwer Academic Publishers,
P.O. Box 17, 3300 AA Dordrecht, The Netherlands.

Kluwer Academic Publishers incorporates the publishing programmes of
D. Reidel, Martinus Nijhoff, Dr W. Junk and MTP Press.

Sold and distributed in the U.S.A. and Canada
by Kluwer Academic Publishers,
101 Philip Drive, Norwell, MA 02061, U.S.A.

In all other countries, sold and distributed
by Kluwer Academic Publishers Group,
P.O. Box 322, 3300 AH Dordrecht, The Netherlands.

Printed on acid-free paper

Printed in the Netherlands

TABLE OF CONTENTS

INTRODUCTION

N.M. VAN STRAALEN** and D.A. KRIVOLUTSKY*
***Department of Ecology and Ecotoxicology*
Vrije Universiteit, De Boelelaan 1087,
1081 HV Amsterdam, The Netherlands
**Institute of Evolutionary Animal Morphology and Ecology*
Russian Academy of Sciences, Leninsky Prospekt 33
117071 Moscow, Russian Federation

Many industrialized and developing countries are faced with the assessment of potential risks associated with contaminated land. A variety of human activities, including municipal waste disposal, industrial emissions, military testing, and agricultural practices have left their impacts on soils in the form of elevated, and locally high concentrations of toxicants. In several cases sources have not yet been stopped and contamination continues.

Decisions on the management of contaminated sites require information on the extent to which toxicants adversely affect the soil ecosystem. For this purpose, it is often insufficient to extrapolate from abiotic sampling. The detection of a toxicant in the abiotic environment usually does not allow a very strong conclusion on the potential hazards. The probability of adverse effects is a function of the physiological state of the organism and of the level of toxicant in the body. The latter again depends on the bioavailable fraction of the chemical in the soil, and this is usually not a fixed fraction of the total amount. The processes involved are so complicated that it seems appropriate to define the hazard not in terms of the concentration present in soil, but in terms of the concentration in organisms. A further step can be made, the hazard of a contamination can be expressed in terms of the observed reduction in essential ecological variables, such as the presence of certain species, the composition of a community, or the rate of dynamic processes. This information can be derived from observations made at the contaminated site itself (*in situ* techniques), or from observations made on samples taken from the site and subjected to testing in the laboratory. All these activities go under the general term bioindication.

A bioindicator system is a tool to extract biological information from an ecosystem and to use that information for making scientifically based decisions on the management of polluted sites. The qualification "system" is used here to emphasize that we are not dealing with a single

N. M. van Straalen and D. A. Krivolutsky (eds.), Bioindicator Systems for Soil Pollution, 1–4.

species acting as a "freeze-dried talking fish on a stick" [1], but with a coherent set of items, each contributing an essential part of the information. With this book we have made an attempt to define such bioindication systems for soil pollution.

Bioindication activities can be conducted on various levels of biological integration. Recent years have seen a proliferation of similar and slightly overlapping terms. The measurement of biochemical and physiological variables in individuals or their excretion products, providing information on exposure or damage, usually goes under the name **biomarker** [2]. The term **bioreporter system** is used for a molecular tool, often using genetically modified cells, to transform the presence of a chemical into an easily measurable signal, such as luminescence [3]. A **biosensor** (also called **bioprobe**), see [4], is a physical device that allows the detection of a chemical as an electric signal derived from a biocatalyst, such as an enzyme or an antibody. A **bioassay** is an ecotoxicological test system, usually of short duration and with a defined protocol, in which the activity of a chemical is measured as an adverse effect on some test species. Finally, all these activities can be repeated in time, this is usually denoted as **biomonitoring** [5]. Following Van Gestel and Van Brummelen [6] we have reserved the term **bioindication** in this book for all activities at the individual (whole organism) level and higher (populations, communities, ecosystems). We pay special attention to systems linking these levels.

The emphasis in this book is on the use of soil animals, in combination with microflora and soil processes. Soil invertebrates live in close contact with the soil and may be expected to reflect soil factors very clearly, Yet, the development of bioindicators that are sufficiently precise to allow the identification of specific soil factors (*cf.* the techniques for biogeochemical prospection using terrestrial plants, see [7]) do not seem to exist for soil animals. This may be due to a general lack of knowledge on the ecophysiological responses of soil invertebrates, combined with the "cryptic" life-style of most of its representatives. This book attempts to contribute to the improvement of this situation and to the stimulation of further research.

The structure of the book is as follows. Part 1 deals with ecotoxicological approaches. The uptake of a chemical by soil organisms is the first step, necessary before any adverse effect can develop. The presence of residues in an organism provides a first indication of bioavailability and possible internal damage. Part 1 gives examples on heavy metals and polycyclic aromatic hydrocarbons. This information must be combined with ecological knowledge on life-cycles and external driving variables, to allow an interpretation of possible risks.

The second part of the book discusses a variety of community and systems approaches. Particular attention is paid to those indicators that use information derived from species with different ecological niches, from different taxonomic groups, and from different scales. In addition, the examples show that it is appropriate to combine information on species and communities with information on dynamic soil processes. There is a need for developing soil bioindication systems to the same degree of sophistication as in the aquatic environment.

In part 3 of the book the various groups of soil invertebrates are discussed with respect to their suitability for bioindication. These are: protozoans, nematodes, earthworms, insects, microarthropods, and macroarthropods, including spiders. The enormous biodiversity of soil life is a necessary prerequisite for the maintenance of the "life-support system" of soils. A solution to the selection problem seems to be that the bioindicators are selected from guilds or functional groups, so that a connection is made with dynamic soil processes, as discussed in part 2.

The fourth part of the book provides some examples of bioindication studies in practice. Case-studies in Russia, USA, Spain and Germany demonstrate the usefulness of bioindication to support the management of contaminated land or the evaluation of waste disposal. The studies cover a variety of contamination problems: radioactive fall-out, heavy metals, PAHs and PCBs. The resources devoted to bioindication programmes vary considerably from one country to another. A hierarchical approach in which determination of community structures is combined with toxicity tests and bioassays, followed by spatial analysis and integrated presentation of data seems to be particularly efficient.

Finally, a chapter of recommendations is added as a summary of discussions held during the workshop that formed the basis for the production of this book. Recommendations emphasize bioindication as a set of coherent activities, the selection of indicators at different levels of organization, and the need to develop indicator systems as doable tools, applicable by regulatory authorities.

References

1. Cairns, J. Jr. (1985) Just give me a freeze dried, talking fish on a stick, *Journal WPCF* **57**, 980.
2. McCarthy, J.F. and Shugart, L.R. (1990) *Biomarkers of Environmental Contamination*, Lewis Publ., Boca Raton.
3. Aarts, J.M.M.J.G., Denison, M.S., De Haan, L.H.J., Schalk, J.A.C., Cox, M.A., and Brouwer, A. (1993) Ah receptor-mediated luciferase expression: a tool for monitoring dioxin-like toxicity, in H. Fiedler, H. Frank, O. Hutzinger, W. Parzefall, A. Riss, and S. Safe (eds.), *Organohalogen Compounds*, Riegelink, Vienna, pp. 361-364.

4. Rawson, D.M. (1993) Bioprobes and biosensors, in P. Calow (ed.), *Handbook of Ecotoxicology* , Blackwell Scientific Publications, Oxford, pp. 428-437.
5. Hopkin, S.P. (1993) *In situ* biological monitoring of pollution in terrestrial and aquatic ecosystems, in P. Calow (ed.), *Handbook of Ecotoxicology*, Blackwell Scientific Publications, Oxford, pp. 397-427.
6. Van Gestel, C.A.M. and Van Brummelen, T.C. (1996) Incorporation of the biomarker concept in ecotoxicology calls for a redefinition of terms, *Ecotoxicology*, in press.
7. Ernst, W.H.O. (1993) Geobotanical and biogeochemical prospecting for heavy metals deposits in Europe and Africa, in B. Markert (ed.), *Plants as Biomonitors. Indicators for Heavy Metals in the Environment*, VCH, Weinheim, pp. 107-126.

Acknowledgments

The editors of this volume are grateful to NATO, Scientific Affairs Division, Brussels, for generously supplying a fund that allowed the organization of an advanced research workshop from which this book is the result. Dr. Ruslan Butovsky and his staff, at the All-Russian Research Institute for Nature Protection in Moscow, made a great effort in the local organization, and significantly contributed to the success of the workshop. A special word of thanks goes to Vera Voznessenskaya, for her consideration of administrative matters. NvS thanks Kees Verhoef and Rotger Rippl for guarding, and Nan Kasanpawiro for support in word processing.

CRITICAL BODY CONCENTRATIONS: THEIR USE IN BIOINDICATION

N.M. VAN STRAALEN
Vrije Universiteit
Department of Ecology and Ecotoxicology
De Boelelaan 1087, 1081 HV Amsterdam
The Netherlands

Abstract

A possible way to evaluate the bioavailability of pollutants in soil is to measure the concentration inside the organisms. This approach is particularly useful in the case of persistent pollutants that are retained by the organism in proportion to the concentration in the environment, and it has been applied often using heavy metals. To allow an estimation of ecological risks from residue data, information is needed on the threshold concentration in the body, above which physiological functions are irreversibly impaired. Lethal body concentrations (LBCs) may be estimated from experiments in which the increase of mortality with exposure time is observed in conjunction with the concentration of metals in the body. LBCs for cadmium have been estimated for several representatives from the soil invertebrate community These data are compared with residue analysis in two forest ecosystems in the Netherlands, to assess and compare the environmental risks of cadmium. It is proposed that the residue - LBC ratio, averaged over a number of species may be used to assess the risks of persistent toxicants at a certain site. The approach should be developed further using internal threshold concentrations for sublethal endpoints, which have a higher ecological relevance.

1. The importance of a toxicokinetic interpretation of residue data

The environmental hazard of a chemical obviously does not depend directly on its concentration in the abiotic environment (soil, water, air). Any adverse effect to the living part of the ecosystem is usually preceeded by uptake into organisms. Although there are some exceptions to this rule (*e.g.* some chemicals causing an irritating effect to the skin

N. M. van Straalen and D. A. Krivolutsky (eds.), Bioindicator Systems for Soil Pollution, 5–16.

without being taken up), the concentration flux into the organism usually provides a better estimate of the chemical's hazard than the external concentration. The flux of a chemical into an animal is, however, difficult to measure, so the researcher must often be content with the measurement of a residue, which is the joint result of uptake and elimination. In many cases the residue is more or less in proportion to the rate of uptake, so it still provides a measure for uptake. In the case of chemicals with a very transient existence in the body (several pesticides), and in the case of chemicals subjected to strong physiological regulation, the residue will not be in direct proportion to the rate of uptake.

The considerations given above define the conditions under which the measurement of concentrations in the body of an animal will give an indication of its bioavailability in soil:

1. It should have a certain persistence in the body; the rate of elimination and metabolization should be in the same order of magnitude as the rate of uptake, resulting in a measurable residue.

2. It should not be regulated by the animal's physiology, in such a way that the internal concentration is independent of the external bioavailability.

Chemicals that meet these conditions are: persistent pesticides, organohalogen compounds, non-essential metals, and polycyclic aromatic hydrocarbons.

Many researchers have followed these lines of reasoning and there is a great amount of data, especially on residues of heavy metals in plants, moss bags, and invertebrates. Data on the terrestrial environment before 1980 have been collected by Martin and Coughtrey [1]. The availability of very sensitive analytical techniques (atomic absorption in a L'vov platform graphite furnace), combined with micromethods for sample preparation [2] has allowed the determination of metals in tiny soil animals, such as individual oribatid mites with a dry body mass down to 20 μg [3, 4].

The data collected for metals show a great amount of inter-species variability, which is obviously related to the physiological mechanisms underlying the storage and excretion of metals [5, 6]. A high residue may be the result of a high rate of assimilation, or a low rate of excretion, or both. Janssen [7] proposed to classify soil invertebrates in five groups, based on laboratory experiments in which uptake and elimination rates were measured for a variety of species. Data summarized in Table 1 illustrate that the measurement of fluxes is more informative than the simple determination of residues, *cf.* [8]. Some groups are relatively homogenous with respect to metal turnover (earthworms), others are very heterogeneous and great differences between species exist (oribatid mites).

TABLE 1. Overview of cadmium turnover in various taxa of soil invertebrates, classified into five groups, based on rates of uptake and elimination, according to [7]

Group	Residue	Assimilation	Excretion
1. Isopods	high	high	low
Snails	high	high	low
2. Millipedes	intermediate	intermediate	?
Centipedes	low	low	?
3. Diplurans	high	?	low
4. Collembola	low	low	high
Carabids	low	intermediate	very high
Ants	low	?	?
Crickets	low	low	high
5. Pseudoscorpions	high	high	low
Harvestmen	high	intermediate	?
Oribatid mites	intermediate/ high	intermediate/ low	low
Gamasid mites	high	high	?
Spiders	high	high	?

The collection of residue data may be useful as it provides an indication of the presence of the chemical and its bioavailability. Another reason to collect residue data is the identification of trends in time and space; many data in the aquatic environment have been collected for this purpose [9, 10]. Conclusions on the risk of a toxicant can, however, not be drawn on the basis of residues only. The species containing the highest residues are not necessarily the ones most at risk. To take the step from residue to risk, some information is needed on the internal concentrations that mark the appearance of adverse effects. This information may be derived from ecotoxicity experiments.

2. The estimation of lethal body concentrations from ecotoxicity experiments

Toxicokinetic compartment models have been very efficient in describing the changes of internal concentrations of chemicals. According to a one-compartment linear model, when test animals are given contaminated food, or are exposed to contaminated substrate, the concentration in the body will increase until an equilibrium is reached between uptake and elimination (Fig. 1). The equilibrium will be directly proportional to the exposure concentration, and if this is high, the internal concentration is likely to approach the lethal body concentration (LBC) after some time. For low external concentrations the internal concentration will equilibrate below the LBC and the animal survives. The dynamic nature of the kinetics causes a decrease in the apparent external LC_{50}, which levels off in proportion to the rate at which the internal concentration reaches its equilibrium. Standard techniques allow explicit equations to be written for the internal body burden, Q, and the LC_{50} as a function of time (Fig. 1).

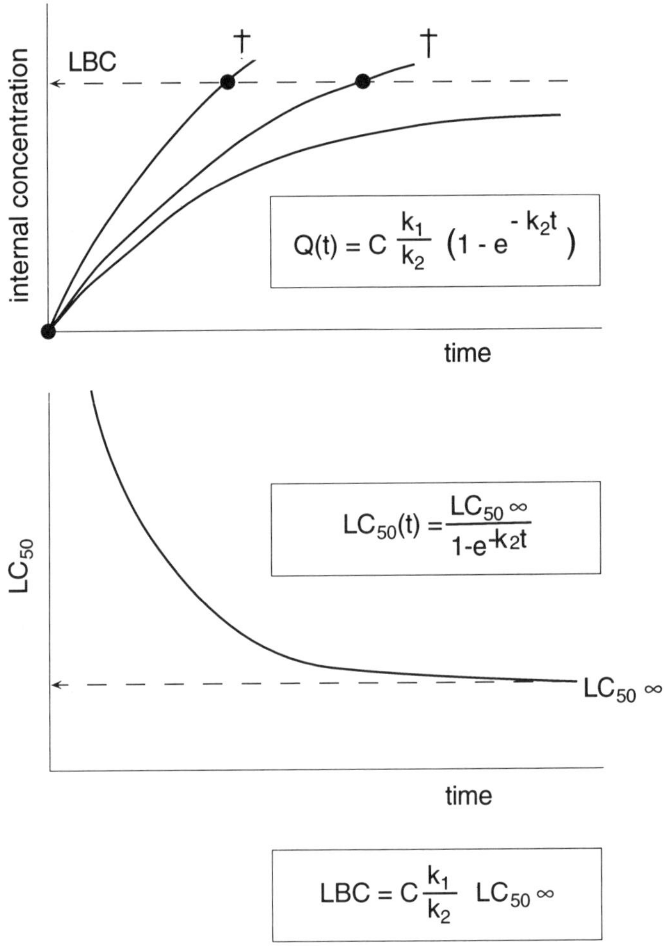

Figure 1. Graphs showing theoretical changes in the internal concentration, Q, and the external LC_{50}, as a function of time, in an ecotoxicity experiment, starting with clean test animals on time zero. The graphs follow the one-compartment toxicokinetic model, with non-zero elimination rate constant.

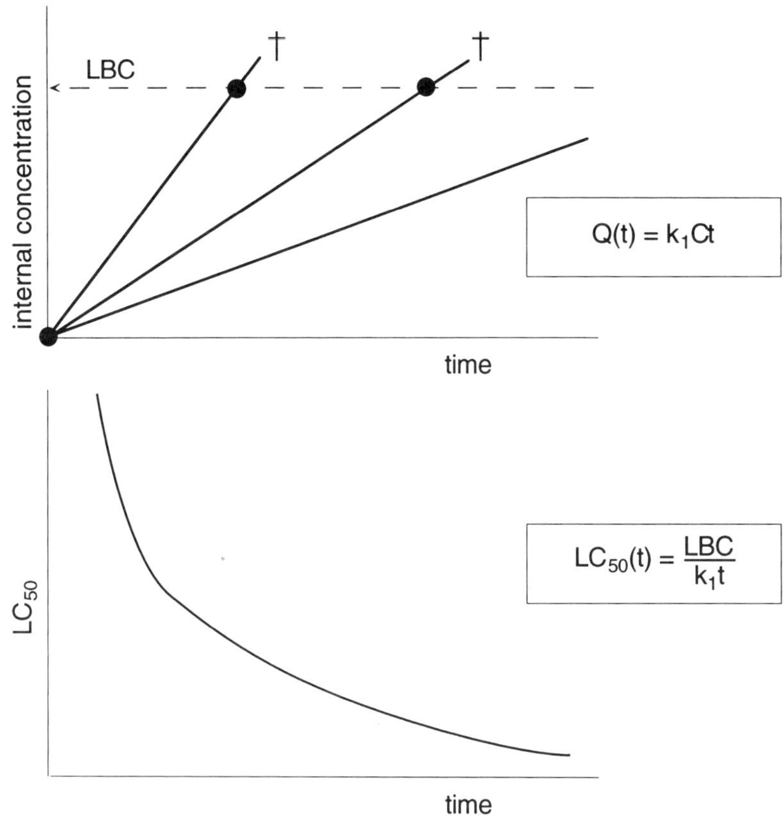

Figure 2. Graphs showing theoretical changes in the internal concentration, Q, and the external LC_{50}, as a function of time, in an ecotoxicity experiment starting with clean animals on time zero. The linear increase of Q, and the hyperbolic decrease of LC_{50} result when the elimination rate constant is zero.

A special case of the model given in Fig. 1 occurs when the elimination rate constant is zero. This is effectively the case for cadmium in isopods, pseudoscorpions, millipedes and diplurans (Table 1). According to the one-compartment model, the internal concentration then increases linearly with exposure time, and the LC_{50} does not approach a constant value, but decreases until zero. The latter prediction is due to the fact that, since every atom of metal is retained in the body, an animal will always reach a lethal body concentration, even with low exposure levels. When the exposure level is very low, the animal will, however, need a very long time to reach the LBC; if this time approaches the natural life-span of the species, the extrapolation to zero LC_{50} looses it meaning.

TABLE 2. Lethal body concentrations of cadmium for some species of soil invertebrates in ecotoxicity-toxicokinetic experiments [14]

Species	LBC (μg/g)
Orchesella cincta	37
Tomocerus minor	75
Folsomia candida	387
Platynothrus peltifer	234
Porcellio scaber	2117
Oniscus asellus	4582
Iulus scandinavius	153

The estimation of LBC from toxicokinetic data combined with ecotoxicity data, may be done using common curve-fitting techniques [11]. For the aquatic environment, LBCs have been estimated for a great variety of substances and species, especially fish [12]. For a series of homologous non-reactive lipophilic compounds, the LBC, expressed in molar units, appears to be relatively constant. This is interpretated as the body burden that causes a general syndrome of membrane damage (minimum toxicity, or non-polar narcosis, [13]). For reactive chemicals lower LBCs are found, while the lowest LBCs are seen for chemicals with a specific mode of action (*e.g.* pesticides).

For soil invertebrates, lethal body concentrations have been estimated for cadmium by Crommentuijn *et al.* [14]. Data available are reproduced in Table 2. It may be seen that LBCs are high for isopods, low for two species of springtails, and intermediate for another springtail, a mite and a millipede. The differences are very large: the isopod *Oniscus asellus* is is able to tolerate an internal body burden which is 120 times higher than the LBC of the springtail *Orchesella cincta*. In addition, there is no similarity of LBCs among taxonomically related species. The data on isopods, including the difference between *P. scaber* and *O. asellus*, are in good agreement with experiments of Hopkin [16].

4. The use of internal threshold concentrations in risk assessment

Internal threshold concentrations may be defined, not only for lethality, but also for sublethal effects. Like LBC is the internal analog of LC_{50},

ITC (internal threshold concentration) is the internal analog of the NEC (no effect concentration). Van Wensem *et al.* [15] argued that ITCs would be more invariant to environmental conditions, in comparison to NECs. The classical problem of bioavailability differences between laboratory and the field clearly holds for external concentrations, because of the use of artificial media, with unnatural speciation of the metal in laboratory experiments. For this reason, the extrapolation of NECs to field conditions is associated with great uncertainties [17]. This argument is less valid for the internal threshold concentration, because it may be assumed that an animal will be adversely affected at a fixed body burden, independent of how that body burden is reached. If this assumption is true, ecotoxicity data may be corrected for the difference in internal concentrations observed in laboratory and field, and soil standards derived from these experiments may be adjusted [15].

Ecotoxicity data for cadmium, for six species of soil invertebrates were reviewed in [15], and from the mean and standard deviation of these data, soil quality criteria were derived, using the distribution-based extrapolation methodology [17, 18]. Using the NECs from laboratory

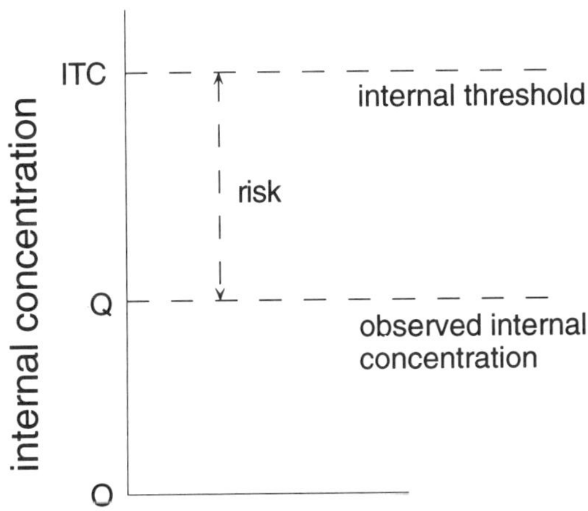

$$\text{risk ratio} = \frac{Q}{ITC} \quad \text{(species A)}$$

$$\beta = \frac{1}{n} \sum_{i=i}^{n} \frac{Q_i}{ITC_i} \quad \text{(community n species)}$$

Figure 3. Schematic representation of the derivation of the "bioindicator index for toxicant residues", β, as the average (taken over a community of n species), of the quotient of the observed residue (Q) in a species, and the internal threshold concentration (ITC) for that species.

tests, a critical level of Cd in soil (hazardous concentration for 5% of the species, HC_5) was estimated as 1.7 µg/g. When corrections were applied for the difference between observed bioaccumulation factors in field studies and in laboratory experiments, the critical level became 0.24 - 0.12 µg/g, depending on whether the bioaccumulation factor was given the lowest or the highest value from an observed range. This exercise showed that biovailability differences between laboratory and field do not necessarily lead to less stringent quality criteria for soils. In this particular example, the variation between species was increased by the corrections, and this effectively decreased the estimate of HC_5.

Another approach towards the use of critical body burdens is to estimate *in situ* risks of metals. This could be done by dividing the measured residue in a certain species in the field by the internal threshold concentration for that particular species (Fig. 3). Preferably, the ratio should be determined for a number of species, and the average value of this ratio is defined here as β, the **bioindicator index for toxicant residues** (Fig. 3). It may be interpreted as the fraction of the internal threshold concentration that is occupied by the residue.

To exemplify this approach, two forest sites in the Netherlands were selected, one being contaminated due to zinc smelter emissions ("Budel"), and one acting as a reference ("Spanderswoud"). Cadmium concentrations have been measured on various occasions in these sites [3, 4, 19, 20]. For three species the data allow a comparison of LBCs and residues, see Table 3. Data in this table again illustrate that the species with the highest residue are not necessarily the ones most at risk. At the Budel site, the highest risk is actually upon the mite *P. peltifer*, which

TABLE 3. Risk assessment of two field sites ("Budel": contaminated, "Spanderswoud": reference), using the "bioindicator index for toxicant residues", β, estimated from cadmium concentrations in three species of soil invertebrates. Q = observed residue [3, 4, 19, 20]; LBC = lethal body concentration (taken from Table 2)

Species	Budel		Spanderswoud	
	Q (µg/g)	Q/LBC	Q (µg/g)	Q/LBC
Porcellio scaber	65.1	0.026	12.0	0.006
Orchesella cincta	1.50	0.041	0.36	0.010
Platynothrus peltifer	15.0	0.064	1.43	0.006
Community		β = 0.044		β = 0.007

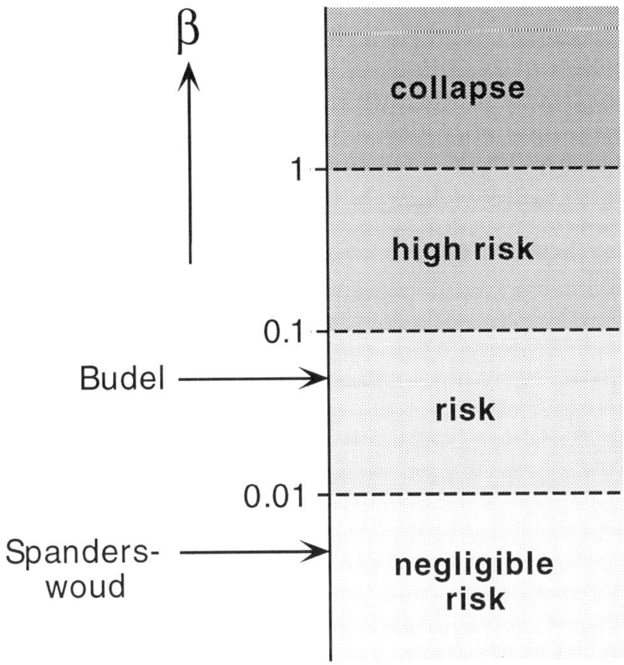

Figure 4. A proposed reference system for evaluating the bioindicator index for toxicant residues, β. The values of β at the two sites of Table 3 are indicated.

has an intermediate cadmium concentration, and not with the isopod *P. scaber*, which has the highest residue, but the lowest risk quotient.

Table 3 also shows that the interspecies variability in observed residues is in the same order of magnitude as the lethal body concentrations. The quotient of Q and LBC does not show such a great variability. On the average, 4.4 % of the LBC is occupied at the Budel site, and 0.7 % at the Spanderswoud site.

To allow an evaluation of the data presented in Table 3, one needs a system of benchmark values for β. A proposal for this is made in Fig. 4. Clearly, the highest value β can take is 1. A value of 1 would mean that, on the average, the community has its LBCs completely occupied; a total collapse of the system is expected above β = 1. The area below 0.01 is tentativily designated as "negligible risk". The benchmark of 1% follows conventions in risk assessment methodology [21]. Experimental work shows that a factor of 100 below the external LC_{50} will be sufficient to protect from sublethal effects [22]. The area between 0.01 and 1 is arbitrarily subdivided into a "high risk" and a "risk" area, using the benchmark value of 0.1 (Fig. 4). Based on this reference system, cadmium concentrations at the site Spanderswoud would be characterized as posing a negligible risk, while the Budel site would be at risk, but not at high risk. Obviously, more comparative data are needed before such a system can be adopted as a standard reference.

4. Limitations of the approach proposed

As stated in the introduction of this chapter, the mere measurement of pollutant residues in organisms does not make sense without some reference system. I have argued here that the lethal body concentration (LBC) could serve as a reference for the observed residue to compare with. This assumes that the LBC may be considered as a constant. Some limitations must be mentioned that qualify this assumption.

1. The LBC may depend on the physiological condition of the animal, that is, an animal in a bad condition (*e.g.* due to low energy reserves) may die from an internal dose that is still tolerated by an animal in a good physiological condition [23].

2. The LBC for one metal may depend on the presence of another metal in the body. For example, cadmium may be less toxic when there is a high body burden of zinc.

3. There may be different thresholds for different parts of the body. This is not necessarily a problem, as long as all parts of the body are in dynamic equilibrium with each other. At high exposure rates, however, uptake in the dosed compartment may be quicker than the rate at which equilibrium can be reached with other compartments, and an overshoot may occur. As a consequence, the lethal dose may depend on how that dose is achieved.

4. The LBC may be subject to selection. High metal conentrations in soil may select for the most tolerant genotypes [24]. When an increase of the LBC is part of the tolerance mechanism, LBCs cannot be compared between populations of the same species living at different sites.

5. The approach assumes that the residue reflects the flux of metals through the body. This is not the case for metals that are subjected to strong regulation. For these metals the residue will be more or less constant over a broad range of exposure levels, and the LBC will be only slightly above the regulated level.

At the moment, there is insufficient information for soil invertebrates to argue which of the above mentioned considerations would seriously invalidate the LBC concept. Further research is obviously necessary, and priority should be given, not to the collection of more residue data, but to the precise estimation of LBCs for more species of the soil invertebrate community. In addition, the approach should be developed further, using internal threshold concentrations for sublethal endpoints (growth, reproduction), which have a higher ecological relevance.

5. References

1. Martin, M.H. and Coughtrey, P.J. (1982) *Biological Monitoring of Heavy Metal Pollution*, Applied Science Publishers, London.
2. Bengtsson, G. and Gunnarsson, T. (1984) A micromethod for the determination of metal ions in biological tissues by furnace atomic absorption spectrophotometry, *Microchem. J.* **29**, 282-287.
3. Van Straalen, N.M. and Van Wensem, J. (1986) Heavy metal content of forest litter arthropods as related to body-size and trophic level, *Environ. Pollut. Ser. A.* **42**, 209-221.
4. Janssen, M.P.M. and Hogervorst, R.F. (1993) Metal accumulation in soil arthropods in relation to micro-nutrients, *Environ. Pollut.* **79**, 181-189.
5. Hopkin, S.P. (1989) *Ecophysiology of Metals in Terrestrial Invertebrates*, Elsevier Applied Science, London.
6. Dallinger, R. (1993) Strategies of metal detoxification in terrestrial invertebrates, in R. Dallinger and P.S. Rainbow (eds.), *Ecotoxicology of Metals in Invertebrates*, Lewis Publ., Boca Raton, pp. 245-289.
7. Janssen, M.P.M. (1991) *Turnover of Cadmium through Soil Arthropods*, Ph.D. thesis, Vrije Universiteit, Amsterdam.
8. Rainbow, P.S. (1988) The significance of trace metal concentrations in decapods, *Symp. zool. Soc. London* **59**, 291-313.
9. Kraak, M.H.S., Scholten, M.C.Th., Peeters, W.H.M., and De Kock, W.Chr. (1991) Biomonitoring of heavy metals in the Western European rivers Rhine and Meuse using the freshwater mussel *Dreissena polymorpha*, *Environ. Pollut.* **74**, 101-114.
10. Hopkin, S.P. (1993) *In situ* biological monitoring of pollution in terrestrial and aquatic ecosystems, in P. Calow (ed.), *Handbook of Ecotoxicology*, Blackwell Scientific Publications, Oxford, pp. 397-427.
11. Janssen, M.P.M., Bruins, A., De Vries, T.H., and Van Straalen, N.M. (1991) Comparison of cadmium kinetics in four soil arthropod species, *Arch. Environ. Contam. Toxicol.* **20**, 305-312.
12. McCarty, L.S. and Mackay, D. (1993) Enhancing ecotoxicological modeling and assessment, *Environ. Sci. Technol.* **27**, 1719-1727.

13. Verhaar, H.J.M. (1995) *Predictive Methods in Aquatic Toxicology*, Ph.D. thesis, University of Utrecht.
14. Crommentuijn, T., Doodeman, C.J.A.M., Doornekamp, A., Van der Pol, J.J.C., Bedaux, J.J.M., and Van Gestel, C.A.M. (1994) Lethal body concentrations and accumulation patterns determine time-dependent toxicity of cadmium in soil arthropods, *Environ. Toxicol. Chem.* **13**, 1781-1789.
15. Van Wensem, J., Vegter, J.J., and Van Straalen, N.M. (1994) Soil quality criteria derived from critical body concentrations of metals in soil invertebrates, *Appl. Soil Ecol.* **1**, 185-191.
16. Hopkin, S.P. (1990) Species-specific differences in the net assimilation of zinc, cadmium, lead, copper and iron by the terrestrial isopods *Oniscus asellus* and *Porcellio scaber*, *J. Appl. Ecol.* **27**, 460-474.
17. Van Straalen, N.M. and Denneman, C.A.J. (1989) Ecotoxicological evaluation of soil quality criteria, *Ecotox. Environ. Safety* **18**, 241-251.
18. Aldenberg, T. and Slob, W. (1993) Confidence limits for hazardous concentrations based on logistically distributed NOEC toxicity data, *Ecotox. Environ. Safety* **25**, 48-63.
19. Janssen, M.P.M. (1988) Species dependent cadmium accumulation by forest litter arthropods, in *Proc. Int. Conf. Environmental Contamination, Venice*, CEP Consultants, Edinburgh, pp. 436-438.
20. Essenin, A.V. and Van Straalen, N.M. (1995) Metal distribution in invertebrates and in some components of pine forests (in the Netherlands) (in Russian), *Toxicol. Rev. (Moscow)* **1995-1**, 20-26.
21. Van Leeuwen, C.J. and Hermens, J.L.M. (1995) *Risk Assessment of Chemicals. An Introduction*, Kluwer Academic Publishers, Dordrecht.
22. Crommentuijn, T., Doodeman, C.J.A.M., Van der Pol, J.J.C., Doornekamp, A., Rademaker, M.C.J., and Van Gestel, C.A.M. (1995) Sublethal sensitivity index as an ecotoxicity parameter measuring energy allocation under toxicant stress: application to cadmium in soil arthropods, *Ecotox. Environ. Safety* **31**, 192-200.
23. Donker, M.H. (1992) Energy reserves and distribution of metals in populations of the isopod *Porcellio scaber* from metal-contaminated sites, *Funct. Ecol.* **6**, 445-454.
24. Posthuma, L. and Van Straalen, N.M. (1993) Heavy-metal adaptation in terrestrial invertebrates: a review of occurrence, genetics, physiology and ecological consequences, *Comp. Biochem. Physiol.* **106C**, 11-38.

CONSTRAINTS IN THE USE OF BIOINDICATORS AND BIOMARKERS IN ECOTOXICOLOGY

P.J. MIGULA
University of Silesia, Department of Human and Animal Physiology
40–007 Katowice, ul. Bankowa 9, Poland

1. Introduction

A good bioindicator system should be based on well recognised and combined methods developed in laboratory and later validated in the field. The use of biomarkers in biological monitoring has been proposed as the new approach [1,2,3]. Their use as the tools for environmental risk assessment has been widely accepted [4,5].

A usefulness of specific biomarkers is limited to the situations when a given pollutant or a group of (closely related) pollutants showing a similar mechanism of action) might be easily recognised and identified quantitatively. Peakall and Walker [6] stated that usefulness of specific biomarkers is reduced only to a given toxic compound, but not providing appropriate information on the presence of any other pollutant. The number of possible contaminants, by–products of man is overwhelming [2], therefore it is important to develop the non–specific biomarkers, which might assess the adverse effects, caused by pollutants treated generally as the stressing factors [7,8,9,10].

Biochemical or physiological biomarkers have been used mainly in laboratory or semilaboratory experiments and further combined with the field observations [9,10]. The links between cause and effect came from laboratory investigations, and then qualitative and quantitative data on a given chemical (chemicals) are derived from the field studies, giving the basis for correlation, necessary for further prediction of the risk for a target group.

Diversity in ecotoxicological responses within the range of organisms has been also generally accepted. Van Straalen [11] has recently proposed a good explanation of interspecific differences in sensitivity and its use in risk assessment. He concluded, that "The action of toxicants on animals cannot be studied without considering all other environmental factors affecting the animal". This statement is in agreement with the concept of multistress used by Nuorteva [12] for the explanation of the forest decline in Europe. In any analysis of interspecific variation at both, the organism and

N. M. van Straalen and D. A. Krivolutsky (eds.), Bioindicator Systems for Soil Pollution, 17–29.

the population level, it is necessary to relate it with changes of important environmental abiotic factors, such as the temperature, water supply, insolation, or the food availability [13,14,15]. Moreover, interactions between pollutants should also be taken into account. This approach stimulates a progress in biomarkers strategy, but has some important limitations, which will be discussed further in details. Three elements, which remain undeveloped, but are important in analysis of ecotoxicological responses of soil animals will be discussed, using examples from the studies on insects and earthworms:

(a) The problem of temperature, as the main factor deciding on metabolism rates in animals, which may exert different effects in various environmental conditions under pollution stress.

(b) Problem of comparability of biological samples. How to compare material in case of social organisms, (e.g. ants) as there may be different response within various social groups or even within the same caste.

(c) Importance of exposure routes in predicting the risk caused by pollutants (e.g. pesticides) to various soil biota compartments.

2. Importance of temperature conditions in bioindication of pollution

The effects of temperature on pollutant kinetics, its bioavailability and toxicity is still underdeveloped [16]. At higher temperatures we would expect lower bioaccumulation of pollutants, since processes of their elimination and detoxication may intensify [15, 17]. Existing data showed the problem is more complicated and should be solved ona species level. Studies of Janssen and Bergema [18] on effects of temperature on cadmium kinetics in collembolans (*Orchesella cincta*) and oribatid mites (*Platynothrus peltifer*) have clearly shown that extrapolation of results from one species to arthropod communities is difficult. The collembolans increased assimilation and excretion of cadmium with temperature, while the mites assimilated higher amounts of metals but no increase was stated for excretion. Dallinger and Wieser [19] indicated that isopods cumulate more copper taken from eaten leaf material at lower temperature since at higher temperatures consumption rates are higher, digestion is less effective and food is retained shorter in the gut.

2.1. TEMPERATURE AND OXYGEN CONSUMPTION AS A BIOMARKER

Oxygen consumption was assumed as one of valuable physiological non–destructive biomarker in environmental assessment made on an organism level [2, 6,7]. As the measure of energy losses in various situations and under pressure of various stressing factors it may be used as a biomarker of exposure and at the same time a biomarker of effect, showing adverse effects, resulted in depletion of the oxygen uptake [15,17,20,21,22,23,24]. Without providing information on the presence of a given pollutant, the uptake of oxygen may be used only as a non–specific biomarker [2].

Temperature effects on oxygen consumption were widely studied and Q_{10} values may vary considerably between species in a wide range of temperature. Studies that combine effects of temperature with a kinetic of a given pollutant are relatively scant. In relation to the soil biota it is worth to mention well developed model of temperature related cadmium kinetics and oxygen consumption in a collembolan *O. cincta* and an oribatid mite *P. peltifer* [18]. Some earlier data are also available for the earthworms [20,24]. More detailed studies were carried out on aboveground insect fauna [15,17,25]. At the same temperature indirect effects of many factors might influence relations between the oxygen uptake and a pollutant. It was shown in experiments in which relations between the oxygen uptake by the pea aphid (*Acyrthosiphon pisum*) and a season, a host plant or density of a colony were studied [26]. Also allometric relations between body weight and metabolism were confirmed only for a given developmental phase [26]. We measured temperature effects on oxygen uptake in hemi– and holometabolic insects under various conditions of gaseous and dust pollution [17,24,25]. Slight stimulatory effects (hormesis) ocurred only when low concentration of one pollutant was applied. Oxygen consumption rates plotted against temperature showed inhibitory effects of used pollutants, which were not proportional to the level of pollution. The effects of toxic substances were expressed more drastically at higher temperatures, with significantly different intercept coefficients of appropriate regression equations [25]. Can we thus expect that the costs of intensified processes of detoxication metabolism rate would increase, when more drastic effects that occurred at higher temperatures suggest rather a reduction of such possibilities? On the other hand, a metabolic cost of mixed–function oxidase induction do not necessarily result in detectable higher metabolic costs than low MFO levels. This was not proved for other detoxifying systems, but it is supposed that a natural selection might favour those systems that are energy efficient [27]. Caterpillars of the satin moth (*Leucoma salicis)* used as a representative of holometabolic insects exposed to the same kind of pollutants, showed a similar pattern of metabolic reaction as a hemimetabolic aphid. In the pupal stage effects of pollutants in relation to temperature practically disappeared. In active adults these effects differed between males and females only in one case. Male moths manifested stimulatory effects at low exposure to SO2 with intensified metabolism rate. Metabolic response of females was similar as that of the pupal stage [17]. These measurements have clearly demonstrated differences of developmental stages in sensitivity to pollutants, confirming higher sensitivity of developing organisms [17, 23, 25, 28]. In the models temperature responses should be planned for implementation. Necessary corrections should be focused on the age structure of a target population and a time scale for development. Necessary corrections for differences caused by temperature related responses to pollutants should respond to the level of exposure and species specificity.

The temperature effects are particularly important when surface or subsurface soil fauna is taken into consideration, because the oscillations of temperature in this part are much higher then in a deep soil. It was expected therefore that use of earthworms in biomonitoring of pollution would not provide special problems related to temperature effects. Analysis of available data did not confirm these expectations. The oxygen uptake measured in *Eisenia foetida* exposed at various temperatures to lead showed also more pronounced effects at higher temperatures, uncorrelated with a concentration of metal [20]. Studies on *Lumbricus terrestris* kept in soil enriched with various concentrations of dust (containing Zn and Pb) taken directly from dust collectors in the zinc smelter, and additionally with zinc and lead showed, that temperature effects caused by pollutants may change in relation to the time of exposure [24]. At the time when uptake and elimination rates reached equilibrium, higher temperature showed no effects, but at low temperature oxygen consumption was strongly reduced (Fig. 1). Temperature significantly modified the concentration effects with a pronounced positive response at higher temperatures. This is in agreement with Janssen and Bergema [18] findings that a species better adapted to cumulate heavy metals respond more intensively to increased temperature than a species presumed to be good metal regulators.

Long–term studies of metabolic changes carried out on *L. terrestris* taken from sampling sites in a pollution gradient from newly constructed large iron plant in Southern Poland showed a progressive depletion of oxygen consumption during 3 consecutive years, but the temperature related pattern of oxygen uptake remained practically the same [29].

2.3. TEMPERATURE AND LIFE–HISTORY APPROACH IN BIOMONITORING

Temperature effects should be taken into account with a special care also in case when life–history parameters of bioindicator species are used to assess the environmental impact of chemicals. For this part I took some data from previously done studies on hemi– and holometabolic insects (aphids and moths) exposed in seminatural conditions to aerial pollutants (SO_2, NO_x, air particles with heavy metals, acid rain) with increased concentration of the same pollutant or increased number of acting pollutants [15,17]. These experiments were carried out in conditions allowing exact application of a given pollutant. In "normal" conditions the fecundity rate of *Acyrthosiphon pisum* is favoured at lower temperatures (15°C and 20°C) and stabilises at higher temperatures [16]. When various pollutants were consecutively added (but at low concentrations) the fecundity rate was even higher at the intermediate temperature of 20°C when compared with groups exposed to only one pollutant in increasing concentrations. Also other life–history parameters appeared to be more sensitive at various temperatures: the mean developmental time necessary for maturation of a female and mean time for survival (LT_{50}). Also the intrinsic rate of increase r_m as the summarising measure of various aspects of life history was temperature dependent.

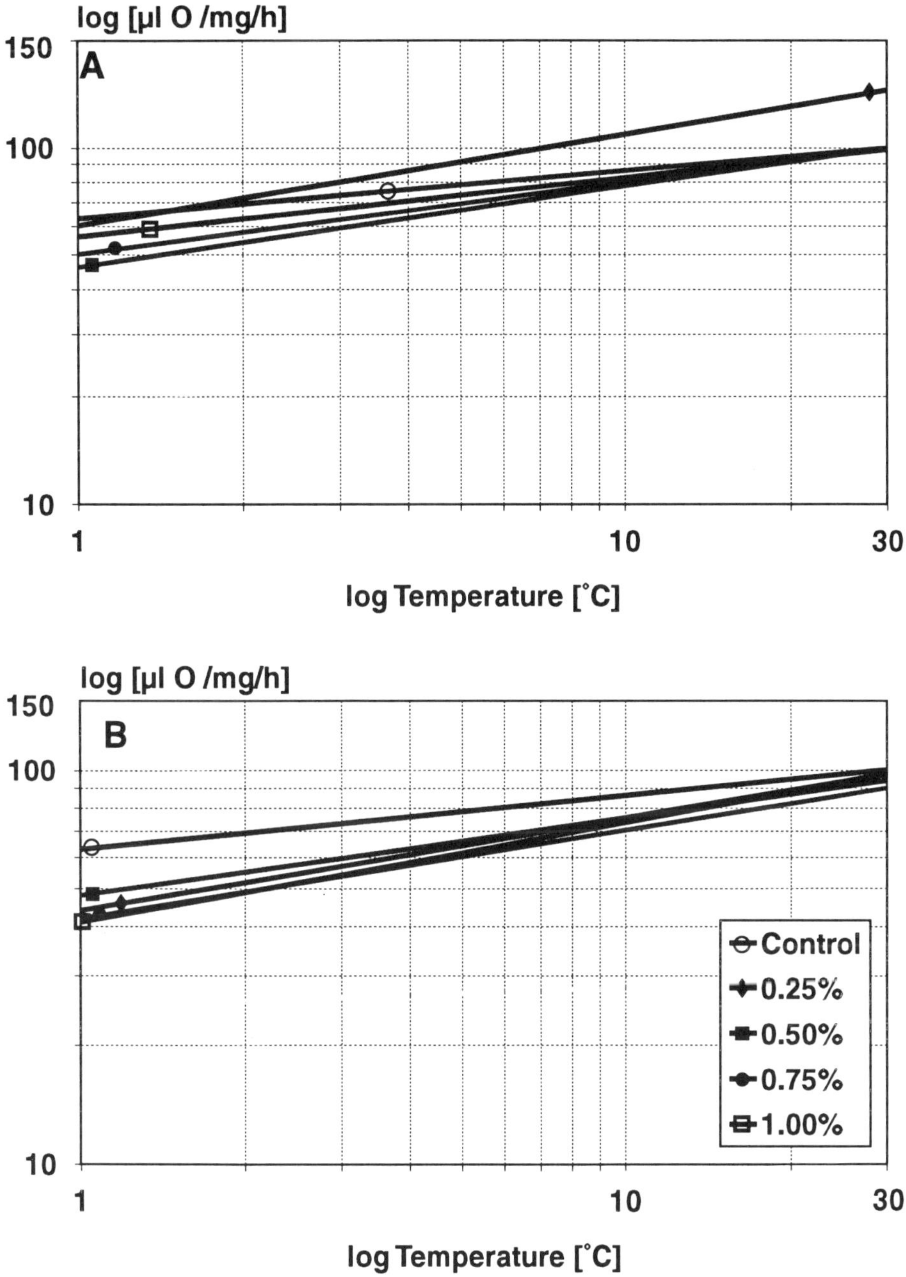

Figure 1. Oxygen consumption by *Lumbricus terrestris* kept for 3 and 10 days in soil with dust containing heavy metals taken from a zinc smelter at different temperatures. Modified from Jethon, et al. [24].

This factor might be modified by developmental time even 10 times more effectively than a fecundity rate of a female [30]. When values of the r_m are experimentally assessed from a laboratory experiment they might be underestimated, since temperature condition are usually kept constant. At fluctuating ambient temperature the r_m may even be higher [31].

Changes of the average temperature threshold for the development may also serve as a good tool for assessing the risk of pollution, but very often a correlation between this factor and a concentration of pollutant is lacking [17].

Concluding, when the life–history approach is applied for risk assessment studies the effects of temperature can not be ignored, since the influence of this abiotic factor may differ when related to developmental or reproductive factors. This approach should be incorporated in further studies examining the relations between pollutants in animals with predominating *"K"* or *"r"* strategy.

3. Constraints in the use of social insects for risk assessment

The role of red wood ants as potential agents of biological control in temperate and boreal forests has been well documented [32,33]. They consume mainly honeydew produced by aphids, but about 10% of their fresh weight food consists of prey animals [32,33,34]. The number of insects killed by a colony during one year has been estimated from five [34] to eight millions [35]. The fate of their colonies is therefore important for functioning of the whole forest ecosystem. The foraging area of the red wood ant family may cover hundreds of square meters, indicating importance of their colonies in estimation of ecological risk from residue data. The red wood ants are known as important accumulators of heavy metals to levels that appear harmful to them [13,36,37,38]. The primary source of metals for ants comes from detritus. Circulation of metals was well described by Nuorteva [13]. Sampling procedures for ants need a special attention [33,38]. Concentration of metals in ants may significantly differ even in the same caste, depending on place where material was collected (Tab. 1).

A comparison of data from various localities showed also seasonal differences between ants taken from the hill and from the foraging routes (Fig. 2). Studies of Finnish ants (*Formica aquilonia*) from unpolluted boreal forest experimentally supplied with cadmium and mercury showed not only the dynamics of distribution of metals within the colony, but also a high ecological vulnerability of these ants, manifested by reduction of the phosphoadenylate pool and disturbances in selected metabolic pathways correlated with body burdens of the metals [37,39,40]. The same model of experiments repeated in moderately contaminated area in Poland showed that a similar critical concentration of cadmium in the body (above 100μg/g dry weight) caused more drastic physiological responses to a metal stress in nested workers than in the foragers, but generally in lower rates than observed in Finnish ants

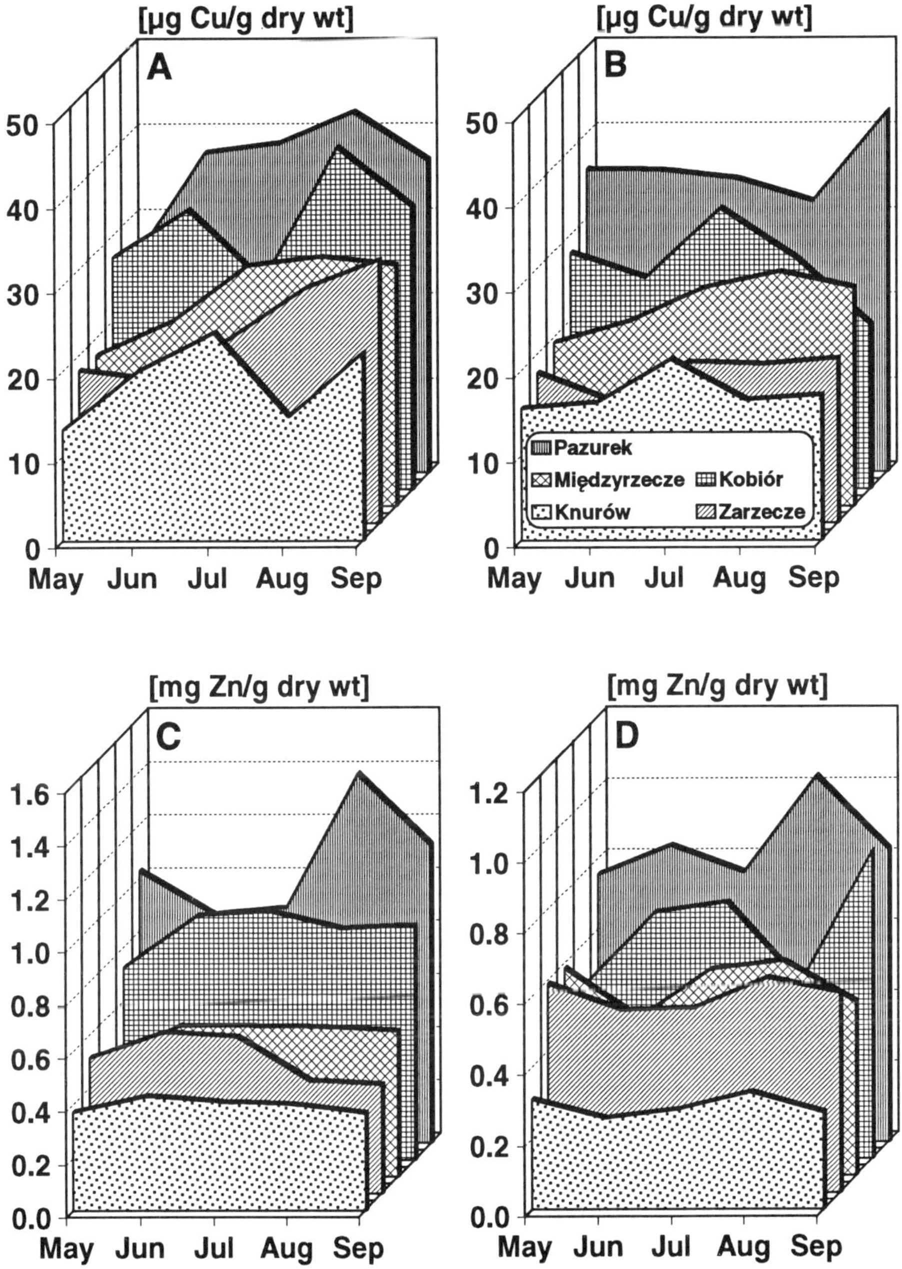

Figure 2. Seasonal changes in the mean concentration of copper and zinc in workers of the red wood ant (*Formica polyctena*):
A, **C** – foraging outside the nest; **B**, **D** – iside the nest [unpublished data]

[38]. The workers support the family with food but the queens decide on gene transfer. The sexuals may be used for comparative studies only during their mating flight.

TABLE 1. Mean concentration of cadmium (μg/ g dry body weight) in various castes and functional groups of workers in the red wood ants (*F. aquilonia* and *F. polyctena)* from variously polluted localities in Estonia [33], Finland [33] and Poland.

Group of ants	Valgeso (Est)	Viikki - (Finl)	Viikki* (Finl)	Pazurek (Pl)	Kobiór (Pl)	Między-rzecze (Pl)	Zarze-cze (Pl)	Knurów (Pl)
Foragers	5.7			45.3	37	21.7	20.7	11.8
Surface workers	5.9	7.1	143.1	51.5	44.7	23.2	16.7	14.4
Inside workers	4.3	6.1	99.6	38.5	27.7	18	12.5	10.7
Pupae	0.1	0.1	6.5	1	0.7	0.6	0.3	0.3
Reserve workers	2.6	3.2	36	29.2	21.1	13.9	9.6	8.5
Males	0.01			3.1	2.2	2.1	1.2	1.8
Queens	0.25	0.5	6.8				7.9	6.3
Larve			6.1	1.4	1.1	0.6	0.7	0.7

* ants from this colony were supplied with 3 x 05. kg honey with 500 ppm $CdCl_2$ [33]

From these studies we may conclude that:
(a) ants may be used as bioindicators for ecological risk assessment only if material for analysis represents various functional groups of workers and winged sexuals
(b) a correlation between metal concentration and various physiological parameters may be invalid if data for comparisons are taken from metal adapted populations or another functional group of workers.

4. Importance of routes of exposure and exposure time in studying effects of pollutants (pesticides)

Identification of potential pollutants and their fate to the soil biota very often has been finished on metals indicated as permanently present compounds. Chemicals such as polynuclear aromatic hydrocarbons (PAH) or pesticides have very short biological half–life but in case of many species have long–term effects [6]. Potential routes, duration and level of exposure to a xenobiotic must be well determined, otherwise a selection of appropriate test to measure possible adverse effects would be impossible [22]. A good procedure needs measurements of various routes of exposure as well as effects of time, with strictly known time of application. Pesticides may appear more toxic to animals living in the upper part of the soil. Calculations of the risk of their exposure may be done only if route and time of exposure is well documented and validated in the field. A good example may give us case studies on honey bees exposed to pyrethroids – a group of insecticides which among other insecticides are probably the most potent, with low mammalian toxicity but high insecticidal efficiency [40]. The half–life of deltamethrin in the soil was calculated approximately for two months [41].

Pyrethroids have been shown highly toxic to pollinators and animals living in the upper part of the soil. The main routes of their entrance are by contact and by consumption of sprayed crop [22]. Using experimentally a series of concentrations as appropriate equivalents of LD_{50} or LC_{50} we have made a comparison of biochemical and physiological responses. Contact exposure generally is based on treatment of an insect with a drop of pyrethroid dissolved in an appropriate solvent as an imitation of spraying – thus a precisely known dose might be used in studying various adverse and time related effects. When insects are kept for a given period on a surface, previously sprayed with a pyrethroid, prediction of concentration effects may be given. The easiest way is to offer a pyrethroid orally – as a dose per insect dissolved in a sugar solution. The analysis of these forms of toxicity indicated strong variations in time–related mortality, which differ in relation to route of the pesticide application and its chemical properties. The mortality curves for deltamethrin were similar after oral and tarsal applications [22]. Reversibly, the oral application of cypermethrin showed a different time–related pattern of action [22]. Similar were patterns of responses of bees after dorsal or oral application of the deltamethrin, but the time of exposure may reveal significantly different effects if metabolism rate is measured on an organism level, isolated tissues or in case of the mitochondrial oxidation [42]. When measurements are done at an individual level the hyperactivity, manifested by a high oxygen uptake, remains at lower concentrations after oral application, but drastically reduces after 24 hours with an increase of the insecticide. After cuticular penetration of a pesticide effects of poisoning may disappear and the recovery is reached after 24 hours (Fig. 3a). At the tissue level (respiratory activity of tissue homogenates) only quantitative differences between both routes of exposure are indicated (Fig. 3b). Dissimilarities were higher when energetic status of the isolated mitochondria is measured (Fig. 3c). Activity patterns of respiratory enzymes or phosphorylases may lead to the similar results [22].

Understanding the way of exposure has of a great importance for differentiation of sensitivity between surface and subsurface feeding species. In case of many groups of

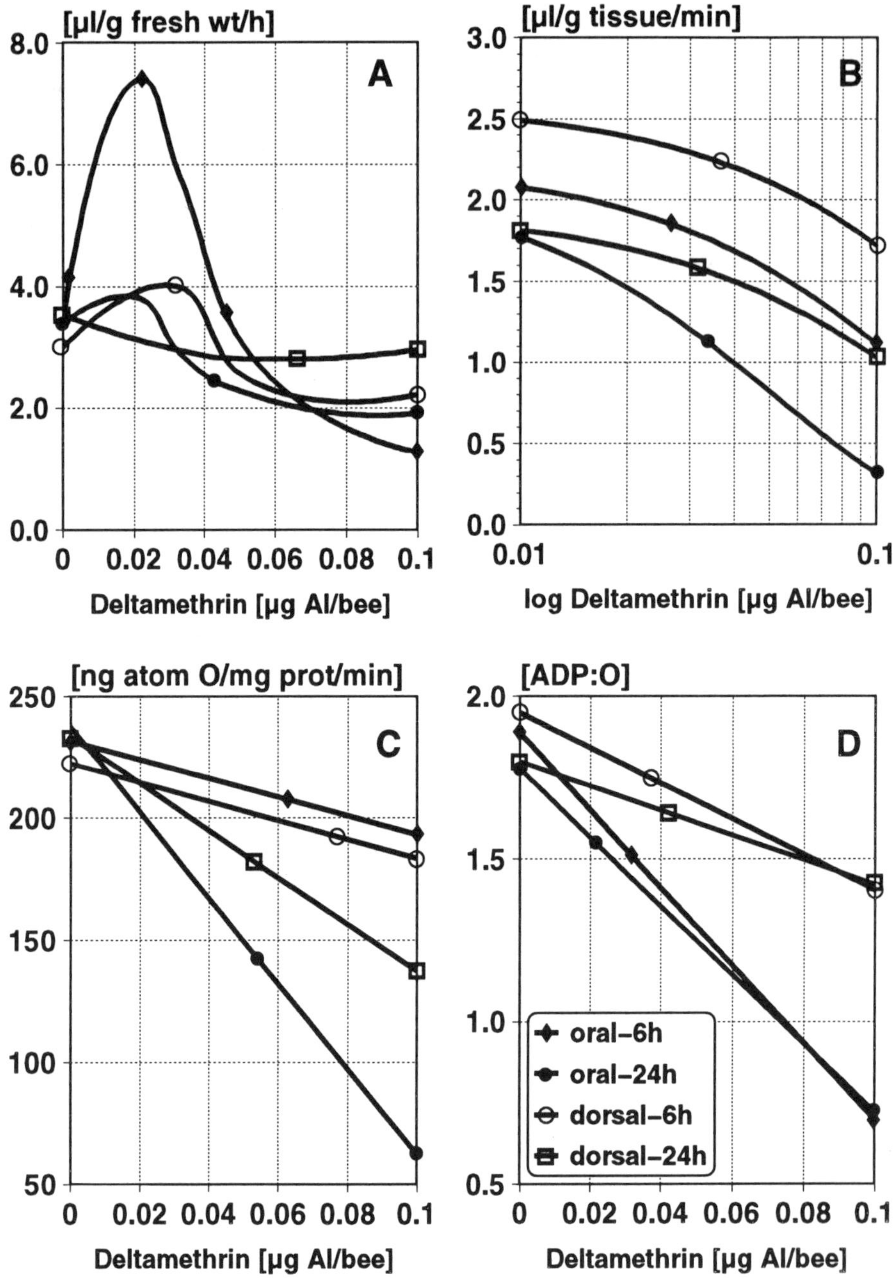

Figure 3. Effect of deltamethrin on metabolism of the honey bee. **A** – oxygen uptake of individuals; **B** – oxygen uptake by tissue homogenates; **C** – State 3 respiration; **D** – Oxidative phosphorylation. Modified [22, 42].

soil animals this approach would be very difficult for implementation, since they may be affected by both routes. For many groups of soil animals our knowledge on feeding habits and mobility patterns is also limited. Only if these elements are estimated, this will improve our methodology of ecological risk assessment.

Acknowledgements

A part of this study related to the red wood ants has been supported by KBN grant No. 0345/P2/93/05

5. References

1. Depledge, M.H. (1993) The rational basis for the use of biomarkers as ecological tools, in M.C. Fossi and C. Leonzio (eds.) *Nondestructive biomarkers in vertebrates,* Lewis Publishers, Boca Raton Fl., pp. 261–285.
2. Depledge, M.H. and Fossi M.C. (1994) The role of biomarkers in environmental assessment (2). Invertebrates, *Ecotoxicology* **3,** 161–172.
3. Fossi, M.C. (1994) Biomarkers in biomonitoring, *Abstr. Third European Conf. on Ecotoxicology,* Zürich, Switzerland, August 28–31.1994.
4. McCarthy, J.F. and Shugart, L.R. (1990) *Biomarkers of Environmental Contamination*, Lewis Publishers, Boca Raton Fl.
5. Peakall, D.B. (1992) *Animal Biomarkers as Pollution Indicators,* Chapman, London.
6. Peakall, D.B. and Walker, C.H. (1994) The role of biomarkers in environmental assessment (3). Vertebrates, *Ecotoxicology* **3,** 173–179.
7. Migula, P. (1989) Bioenergetic indices as indicators of environmental contamination in insects, in S.C. Goel (ed.), *Nutritional Ecology of Insects and Environment,* Muzaffarnagar, India, pp. 157–166.
8. You, S.J. (1983) Induction of detoxifying enzymes by allelochemicals and host plants in the fall armyworm, *Pestic. Biochem. Physiol.* **19**, 330–336.
9. Migula, P. and Głowacka, E. (1995) Heavy metals as stressing factors in the red wood ants (*Formica polyctena)* from industrially polluted forests, *Fresenius J. Chemistry* (submitted)
10. Migula, P., Bińkowska, K., Kafel, A., Kędziorski, A. and Nakonieczny, M. (1989) Heavy metal contents and adenylate energy charge in insects from industrialized region as indices of environmental stress, in J. Bohac and V. Ruzicka (eds.), *Bioindicatores Deteriorisationis Regionis,* Česke Budejovice, **2,** 340–349.
11. Van Straalen, N.M. (1994) Biodiversity of ecotoxicological responses in animals, *Netherlands J. Zoology,* **44,** 112–129.
12. Nuoprteva, P. (1990). *Metal Distribution Patterns and Forest Decline. Seeking Achilles' Heels for Metals in Finnish Forest Biocoenoses,* Univ. of Helsinki Publ. No. 11, Helsinki.
13. Alstad, D.N., Edmunds, G.F. and Weinstein, L.H. (1982) Effects of pollutants on insect populations, *Ann. Rev. Entomol.,* **27**, 369–384.
14. Joosse, E.N.G. and Verhoef, H.A. (1987) Developments in ecophysiological research in soil invertebrates, *Adv. Ecol. Res.,* **16**, 175–248.

15. Migula, P. aand Jethon, Z. (1990) The resistance of aphids and moths to air pollution, *Arch. Ochrony Środowiska,* Zabrze, **3–4**, 141–156.

16. Barron, M.G. (1990) Predictive toxicokinetic models, *Environ. Toxicol. Chem.* **9**, 131–132.

17. Migula, P. (1985) *Wrażliwość wybranych gatunków owadów na skażenia powietrza gazami i pyłami przemysłowymi oraz tolerancja przez nie termicznych zmian środowiska,* Prac. Nauk. No 765, Uniwersytet Śląski, Katowice

18. Janssen, M.P.M and Bergema, W.F. (1991) The effect of tempwerature on cadmium kinetics and oxygen consumption in soil arthropods, *Environ. Toxicol. Chem.* **10**, 1493–1501.

19. Dallinger, R. and Wieser, W (1977) The flow of copper through a terrestrial food chain. I. Copper and nutrition in isopods, *Oecologia* **30**, 253–264.

20. Migula, P., Baczkowski, G. and Wielgus–Serafińska, E. (1977) Respiratory metabolism and haemoglobin concentration in an earthworm – *Eisenia foetida* Savigny 1826, under exposure of lead intoxication. *Acta Biologica,* Katowice, **4**, 64–78.

21. Migula, P. (1979) The effect of lead upon the metabolism rate by larval stages of *Euproctis chrysorrhoea* L., and *Malacosoma neustria* L (*Lepidoptera*), *Acta Biologica,* Katowice, **6**, 63–73.

22. Migula, P., Hurny, J., Kędziorski, A., Nakonieczny, M., Kafel, A. and Binkowska, K. (1990) Metabolic effects of the pyrethroid action in the honey bee, *Uttar Pradesh J. Zoology* **10**, 1–10.

23. Migula, P. (1989) Combined and separate effects of cadmium, lead and zinc on respiratory metabolism during the last larval stadium of the house cricket (*Acheta domestica* L.), *Biologia,* Bratislava, **44**, 513–521.

24. Jethon, Z., Migula, P., Łuszczak, G. and Kędzierska, M. (1987) The effect of soil contamination by dust containing heavy metals on the respiratory metabolism of *Lumbricus terrestris* L., *Biologia,* Bratislava 42, 537–546.

25. Krawczyk, A. (1985) Wrażliwość mszyc na dwutlenek siarki. Cz. I–III, *Acta Biologica,* Katowice, **17**, 103–141.

26. Migula, P. (1986) Metabolism rate in two biotypes of the pea aphid, *Acyrthosiphon pisum* (Harris) (*Homoptera: Aphididae*), *Acta Biologica Silesiana,* Katowice, **5**, 21–33.

27. Neal, J.J. (1987) Metabolic costs of mixed–function oxidase induction in *Heliothis zea, Entomol. exp. appl.* **43**, 175–179.

28. Heliövaara, K. and Väisänen, R. (1993) *Insects and Pollution,* CRC Press, Boca Raton Fl.

29. Pytasz, M., Migula P. and Krawczyk K. (1980). Wpływ emisji przemysłowych na metabolizm ogólny wybranych gatunków zwierząt z rejonu Huty Katowice, *Acta Biologica,* Katowice, **8**, 81–103.

30. Lewontin, R.C. (1965) Selection for colonizing stability, in H. Baker and G.L. Stebbins (eds.), *The Genetics of Colonizing Species*, Academic Press, New York, pp. 77–94.

31. Root, B. and Olson, A.M. (1969) Population increase of the cabbage aphid, *Brevicoryne brassicae,* on different host plants, *Can. Ent.,* **101**, 768–773.

32. Whittaker, J.B. (1991) Effects of ants on temperate woodland trees, in C.R. Huxley and D.F. Cutler (eds.) *Ant – Plant Interactions,* Oxford Univ. Press, pp. 67–79.

33. Maavara, V., Martin, A.J., Oja, A. and Nuorteva P. (1994) Sampling of different social categories of red wood ants (*Formica* s.str.) for biomonitoring, in B. Markert (ed.) *Environmental Sampling for Trace Analysis,* VCH, Weinheim, **26**, 465–489.

34. Rosengreen, R. and Sundström, L. (1991) The interaction between red wood ants, *Cinara* aphids, and pines. A ghost of mutualism past? , in: C.R. Huxley and D.F. Cutler (eds.) *Ant – Plant Interactions,* Oxford Univ. Press, pp. 80–91.

35. Wellenstein, G. (1954) Die insektenjagel der Roten Waldameise, *Z. Angew. Entomol.* **36**, 185–217.

36. Ylä–Mononen, L., Salminen, P., Wuorenrinne, H., Tulisalo, E. and Nuorteva P. (1989) Levels of Fe, Al, Zn and Cd in *Formica aquilonia, F. polyctena* and *Myrmica rufinodis* (Hymenoptera, Formicidae) collected in the vicinity of spruces showing different degrees of needle–loss, *Ann. Entomol. Fennici* **55**, 57–61.

37. Migula, P., Nuorteva, P., Nuorteva, S–L., Głowacka, E. and Oja A. (1993) Physiological disturbances in ants (*Formica aquilonia*) from excess of cadmium and mercury in a Finnish forest, *Sci. Total Environment* Supplement 1993, 1305–1314.

38. Migula, P., Głowacka, E., Nuorteva, S–L., Nuorteva, P. and Tulisalo, E. (1994) Tolerance of cadmium and mercury by the red wood ants (*F. aquilonia*) as possible stressing factors in a Finnish forest. *Abstr. Third European Conf. on Ecotoxicology*, Zurich, Switzerland, August 28–31.1994., pp. 2.12.

39. Nuorteva, P., Nuorteva, S–L., Oja, A., Lehtionen, H. and Salo, S. (1992). Two Achilles heels for metals in the Finnish forest ecosystem, in P. Bohac (ed.), Proc.VIth Int. Conference *Bioindicatores Deteriorisationis Regionsis,* Česke Budejovice, pp. 72–77.

40. Smith, T.M. and Stratton G.W. (1986) Effects of synthetic pyrethroid insecticides on nontarget organisms, *Residue Reviews* **98**, 93–120.

41. Chapman, R.A. and Harris, C.R. (1981). Persistence of four synthetic pyrethroid insecticides in a mineral and organic soil, *J. Environ. Sci. Health* **B16**, 6056–615.

42. Migula, P. (1990) Mitochondrial substrate oxidation in thoracic muscles of the honey bee (*Apis mellifica* L.) treated with deltamethrin, *Acta Biologica Silesiana*, Katowice **15**, 107–117.

POLYCYCLIC AROMATIC HYDROCARBONS IN SOIL DETRITIVORES

Specific accumulation and effects

Jack H. FABER and Guido J.S.M. HEIJMANS
DLO/Institute for Forestry and Nature Research
P.O. Box 23, NL-6700 AA Wageningen, The Netherlands

Abstract

The distribution and effects of polycyclic aromatic hydrocarbons (PAH) were investigated in woodland ecosystems of river floodplains along the river Waal in The Netherlands, studying an inundation gradient in the field, and experimentally exposing soil animals in the laboratory. The field study was aimed at describing PAH levels and profiles in soil detritivore communities in relation to their food resources. A high inundation frequency was reflected by high total PAH contents in soil, but did not affect the average total PAH concentration in soil animals. Within sites, differences up to 2 orders of magnitude could be found between species. It was concluded that PAH compounds were not always accumulated by the fauna, when related to total body weight. If expressed per unit fat content, average bioconcentration factors at the sites ranged between 33-315.

Wood consumers tended to show higher concentrations of total PAH on a fat basis, than did leaf litter feeders and indifferent feeders. Lower PAH compounds were most readily taken up by all groups. Leaf litter feeders as a group showed a profile that was dominated by fluoranthene, whereas wood consumers more often showed highest concentrations of phenanthrene.

Females of the woodlouse *Trachelipus rathkei*, exposed in the laboratory to phenanthrene at 20 or 200 $\mu g.g^{-1}$ food, showed retarded brood development, an increased total number of eggs and mancas, and an increased reproductive effort in

N. M. van Straalen and D. A. Krivolutsky (eds.), Bioindicator Systems for Soil Pollution, 31–43.

terms of biomass allocation by the female. Nett fecundity expressed as the number of live mancas was highest in the 20 $\mu g.g^{-1}$ treatment, as was the percentage of females with succesfull brood.

1. Introduction

1.1. FRAMEWORK

In the Dutch 'Nature Policy Plan' [4] direct and indirect effects of pollution are considered as part of a complex of factors causing the present decline in flora and fauna. The establishment of nature oriented standards, aimed at protection of species and communities, is seen as an important management tool. The current set of standards for environmental quality, however, was considered not fully appropriate for nature management since insufficient consideration is given for bioavailability, effects at the population level, and bioaccumulation and secondairy poisoning in foodwebs. The present studies are part of an extensive research program aimed at setting such nature-oriented standards for contaminants (NSC) of ecosystems.

The program is focussed on polycyclic aromatic hydrocarbons (PAH) as model pollutants, since little ecotoxicological information is available on this widespread group of contaminants. In particular, effects of PAH on ecosystems in floodplains of the rivers Rhine and Waal in The Netherlands are studied. These floodplains make part of a series of wetlands where national and provincial management is aimed at the initiation and further development of natural ecosystems ("Plan Stork"). One approach is to increase inundation dynamics by cutting through summerdikes. This is expected, however, to result in increased sedimentation of polluted river sediments in the floodplains, as well as a redistribution of older and more heavily polluted sediment soils, with consequent potential for adverse effects on the ecosystem. Therefore, in the NSC program bioavailability of PAH in the field is also studied.

1.2. SCIENTIFIC BACKGROUND

Although little is known about PAH in soil invertebrates, the available data indicate that body burdens of total PAH and of specific compounds vary greatly with species [7, 11]. No general ecotoxicological theory (foodchain effects, body size relations) seems applicable, however, to explain these interspecific differences, and the levels

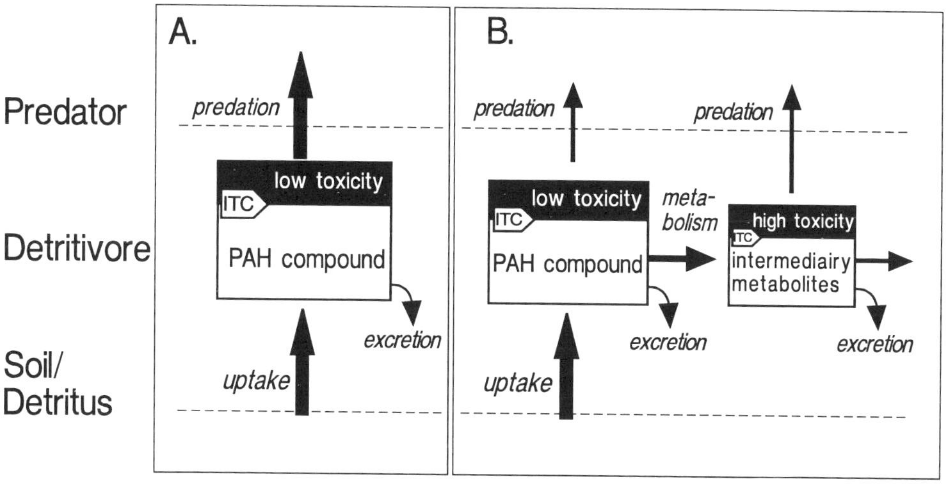

Figure 1. Scheme of routes of PAH through soil detritivore foodchains. Relative size of boxes and arrows indicates PAH pools and fluxes. **A**: Accumulation by detritivore species with no or little metabolisation of PAH and few toxic effects, a potentially hazardous route for predators; **B**: No accumulation by detritivore species with ample metabolization of PAH and potential for severe toxicity from intermediairy metabolites, a less hazardous route for predators. ITC, internal threshold concentration.

of PAH more likely may be affected by species-specific characteristics such as uptake and excretion efficiencies, capability for biotransformation, and fat content [1]. The large taxonomic diversity of the soil community should facilitate further investigations of factors determining the distribution of PAH in biota.

Terrestrial invertebrates in general were considered to have little ability for metabolization of aromatic compounds, apparent exceptions were attributed to gut micro-organisms [3]. Biotransformation may occur in some species, however, e.g. the isopod *Porcellio scaber* [Van Brummelen, pers. comm.], and general or specific mixed function oxidase enzyme activities have been reported for various taxonomical groups, e.g. lumbricids [3], insects [5], and molluscs [15]. Soil detritivores in particular might turn out to be amongst those, as these are often confronted with complex plant chemicals through their food.

If a detritivorous species is incapable of biotransformation, it supposedly might be more liable to show bioaccumulation when confronted with PAH. Predators of such species may then be at a relative enhanced risk for secondary poisoning (Fig. 1A). On the other hand, detritivores with potential for biotransformation of PAH may not endanger their predators so much, but rather suffer themselves from the higher toxicity of intermediairy metabolites (Fig. 1B). Thus, various PAH compounds have been demonstrated to affect growth, reproduction or survival in earthworms [1, 6], collembola [1], and isopods [10, 12].

1.3. WHY SYSTEM IS NEEDED

An inventory of PAH pollutants in terrestrial biota is needed as basis for recognition of hazardous foodchains that find their basis in the soil community. One: as relative concentrations of various PAH compounds may vary with the source of contamination [2, 9], an inventory of PAH profiles in soil animals in relation to soil concentrations may reveal species that preferentially take up and accumulate specific PAH compounds. Thus, a means of bioindication for specific types of pollution may be developed. Second: an inventory of PAH in detritivorous prey species is needed in order to discriminate what feeding types of vertebrate predators, particularly mammalian and avian species, may be at risk at a particular site.

Direct toxic effects of PAH on soil detritivores should also be studied as this group is also functionally important in decomposition and pedogenesis processes.

2. Proposed new system

In this paper a preliminary overview is presented of PAH contents in a broad taxonomical variety of soil detritivorous species in river floodplain woodland ecosystems. The data are analysed with respect to detailed feeding habits of these species, discriminating between wood feeders, litter consumers, and indifferent species, consuming both food types. These feeding guilds can be considered to be more or less separated in view of microhabitat, whilst being specifically exposed to PAH through their food resources, with each having specific predators.

Further, some of these species are tested in the laboratory for their sensitivity towards various PAH compounds. In this paper results are presented on some life-history parameters of the isopod *Trachelipus rathkei* as affected by phenanthrene.

3. Example of application

The feeding guild approach is tested on data on PAH in terrestrial detritivores with an epigeic lifestyle. The data were collected as part of a small monitoring study within the framework of the NSC research program.

3.1. MATERIALS AND METHODS

3.1.1. *Sampling sites*

Various soil detritivores, all macrofauna species, were collected by hand-sorting at three floodplain sites along the river Waal, The Netherlands, in December 1994. The vegetation encompassed open to closed woodland, dominated by *Populus* or *Salix* species. The sites represented different frequencies of inundation and degrees of soil pollution with PAH (Table 1):

(a) 'seldom', a site with a low degree of PAH pollution, rarely inundated (last inundation occurring in 1923), 18 soil fauna species (6 taxonomic orders);

(b) 'infrequent', with inundation every few years and moderate levels of pollution, 3 species (3 orders);

(c) 'frequent', with yearly inundation and with relatively high levels of pollution, 6 species (5 orders).

3.1.2. *Sample treatment and analysis*

Upon collection, animals were kept in the laboratory at 4 °C on moist filterpaper for defaecation for 2-6 days, until the guts had emptied. Prior to analysis animals were cleaned from adhering debris under a dissecting microscope. PAH analyses were performed in acetonitril using HPLC, with a ChromspherPAH column and

TABLE 1. Characteristics of sampling sites.

Site inundation	Inundation frequency *(yr^{-1})*	Soil type	PAH contents[#] *($mg.g^{-1}$)*	Organic matter *(%)*
Seldom	0.02	sand	2.41[*]	4.7[*]
Infrequent	0.5	clay	4.97	0.4
Frequent	1	clay	25.39	2.4

[#] PAH contents standardized for soil with 10% organic matter.

[*] determined after an inundation in December 1994.

fluorescence detection, after soxhlet extraction in hexane and concentration of samples in Kuderna Danish evaporators, and purification over aluminumoxide. Fat contents of samples were determined gravimetrically in unpurified extracts.

Soil samples (top 5 cm) were extracted with acetone/petroleum ether, leaf litter and woody materials were soxhlet extracted in hexane. The extracts were subsequently processed as described above. Organic matter fractions were determined as loss on ignition after 16 hours at 375 °C.

3.2. RESULTS AND DISCUSSION

Whilst total PAH levels in soil were highest at the frequently inundated site, total contents of the soil detritivorous fauna, averaged over all species, showed little variation amongst sites (Table 2). This indicates that bioavailability of PAH at the most polluted site is relatively low. Indeed, a large portion of total PAH is immobilized by coal particles in the river sediment [Wieggers, pers. comm.], thus being protected from root uptake and microbial breakdown, once deposited. No bioconcentration of total PAH by the soil fauna was found when expressed on the basis of body weight, but on a fat basis the average BCF_{soil} rated 33 under frequent inundation, and, contrastingly, 315 at the seldomly inundated site. Since between species differences in total PAH concentrations varied by two orders of magnitude (not shown), specific BCF values varied concordingly. Thus, bioconcentration of individual species ranged between 10 (a mollusc, cf. *Arianta spec.*) and 1360 (a mixture of enchytraeids, 98% *Fridericia ratzeli*), both extremes occurring at one site ('inundation seldom'). Body burdens were generally two times lower than described for a blast furnace area [11], and equalling data reported for an urban roadway system [7]. In the latter study animal guts were not cleared, though, therefore these data may represent overestimates.

Concentrations of individual PAH compounds in animal tissues showed only small differences between sites (Figure 2), alike concentrations of total PAH. Only fluorene showed a higher concentration at the frequently inundated site. It was also found (Table 2) that lower compounds dominated PAH profiles at all sites, both in soils and animal tissues. However, animal PAH profiles typically were more dominated by these compounds than were soils, particularly by fluorene, phenanthrene and fluoranthene. Litter and wood showed intermediate PAH profiles, indicating that aerial input into the system was insignificant, elsewise other compounds, characterizing a different source, would co-dominate [9].

TABLE 2. Profiles of PAH compounds in soil and soil detritivores relative to benzo(a)pyrene (BaP). The actual concentrations of BaP and total PAH are also given (μg/g soil, standardized to 10% organic matter, or ng/g fresh wt. for leaf litter, wood, and animal samples).

Site	Sample	PAH compound														BaP	totPAH
inundation		ACE	FLU	FEN	ANT	FLA	PYR	BaA	CHR	BbF	BkF	BaP	DBA	BgP	IP		
Frequent	Soil	<0.19	0.1	0.8	0.3	1.9	1.5	1.1	1.1	1.0	0.5	1.0	0.2	0.7	<	2.48	25.39
	Fauna	0.8	4.2	4.8	0.1	4.2	3.6	1.0	2.3	1.7	0.7	1.0	<	1.4	<	2.38	50.66
Infrequent	Soil	0.1	0.1	0.9	0.2	1.6	1.3	0.9	1.0	1.2	0.5	1.0	0.2	0.8	<	0.48	4.97
	Fauna	<	2.6	11.1	0.4	10.1	6.5	1.5	5.2	2.3	0.9	1.0	<	1.1	<	1.21	49.79
Seldom	Soil	<	0.1	1.0	0.3	2.0	1.5	1.0	1.0	1.5	0.5	1.0	0.2	1.0	1.0	0.25	2.41
	Fauna	1.3	4.6	12.0	0.4	12.2	6.4	1.5	4.7	2.0	0.8	1.0	1.2	1.4	<	1.12	44.09
	leaf litter	0.4	1.4	16.0	0.2	8.8	6.2	2.4	3.6	2.3	1.1	1.0	7.3	1.2	1.6	6.81	303.65
	wood	<0.1	1.7	6.5	0.1	4.8	2.0	0.9	1.5	1.5	0.7	1.0	11.8	1.1	0.8	5.19	113.29

< Below detection limit

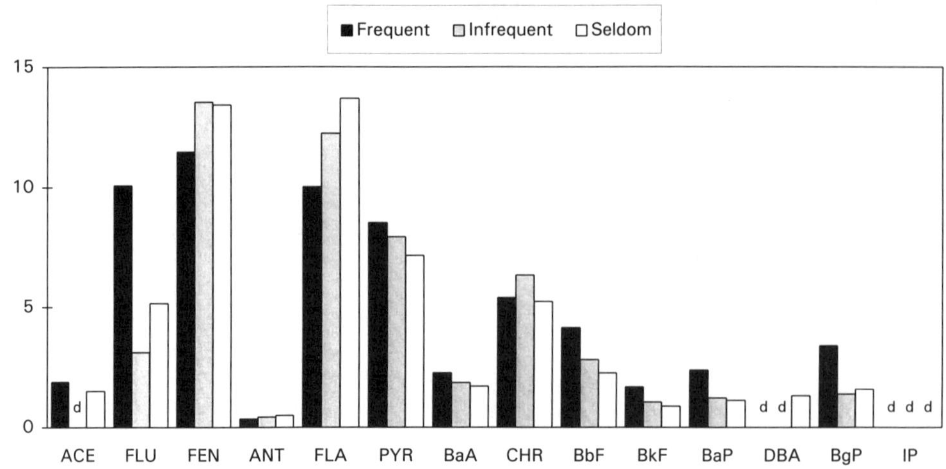

Figure 2. Average PAH contents (in $ng.g^{-1}$ body wt) of soil detritivores at river floodplain sites with various inundation frequencies and pollution levels. Abbreviations: ACE, acenaphtene; FLU, fluorene; FEN, phenanthrene; ANT, anthracene; FLA, fluoranthene, PYR, pyrene; BaA, benzo(a)anthracene; CHR, chrysene; BbF, benzo(b)fluoranthene; BaP, benzo(a)pyrene; DBA, dibenzo(a,h)anthracene; BgP, benzo(g,h,i)perylene; IP, indeno(1,2,3-c,d)pyrene; d, below detection limit.

The specific differences in bioconcentration rates could not be related to variation between higher taxa. The data for total PAH body burden were then analysed in view of feeding habits of individual species, as based on literature data and field observations. Three feeding guilds were recognized: wood consumers, leaf litter consumers, and indifferent species. PAH concentrations were pooled for all species at each sampling site, and relative proportions were calculated for each guild. The results, summarized in Figure 3, indicate that consumers of wood generally accumulate higher levels of PAH than consumers of litter or indifferent species. At the most polluted site, this relative difference was most evident. Surprisingly, indifferent species did not occupy an intermediate position between wood and litter consumers. At the species level these observations were not consistent, however, and confirmation is needed from further studies.

Wood consumers frequently showed bioconcentration factors greater than unity (BCF_{food} based on total body wt), whereas leaf litter consumers did not. On a fat basis, bioconcentration to food generally was 10-fold higher than to soil.

TABLE 3. Specific accumulation of PAH compounds by different soil detritivore feeding guilds. Average weighed frequency of top 4 dominating compounds. For abbreviations see legend to Fig. 2.

Feeding guild	PAH compound					
	FLU	FEN	FLA	PYR	CHR	BbF
Leaf litter consumers	0.44	1.82	2.47	1.41	1.38	0.18
Wood consumers	0.28	5.97	2.63	1.12	1.67	0
Indifferent species	1.11	4.59	3.48	1.98	1.11	0.23

Some specific differences were observed in accumulation levels of individual PAH compounds. These again could not be ascribed to taxonomical groups. The data were then analysed in view of feeding guilds, after ranking of the top 4 PAH compounds in every species, and weighing of average frequencies per group within sites. As shown in Table 3, PAH profiles of wood feeding species most often were dominated by phenanthrene, whereas in the litter consuming guild fluoranthene was dominant in most cases. Indifferent species showed intermediate distributions of PAH compounds. Within the guild of wood consumers differences were detected between species in accumulation rate of individual PAH compounds (not shown). March fly larvae, *Bibio marci* (L.), predominantly accumulated high molecular compounds (mol. wt. $\geq$ 228.3), while the enchytraeid *Fridericia ratzeli* (98% of individuals in sample) contrarily showed accumulation of low molecular compounds.

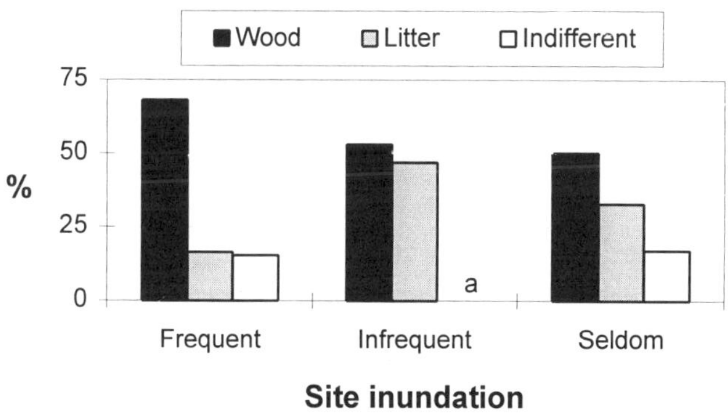

Figure 3. Relative PAH contents (expressed as percentage of the total quantitity of PAH, as determined in individual species body fat) in detritivore feeding guilds at river floodplain sites with various inundation frequencies and pollution levels. a; sample material insufficient for analysis.

4. Effects of PAH

In addition to the above field studies, effects of PAH on invertebrate growth and reproduction were studied in the laboratory. The isopod *Trachelipus rathkei* Brandt, a dominant species at all sites, was chosen as a test species. PAH concentrations in the field were above the average for soil detritivores, and ranged between 510-726 $\mu g.g^{-1}$ fresh wt. Since the ability to metabolize PAH has been demonstrated *in vitro* in isopods [Van Brummelen, pers. comm.], it can be hypothesized that this species is succeptable to intoxication. In this paper some results of a study on effects of phenantrene are reported.

4.1. MATERIAL AND METHODS

Females were captured at the site 'inundation seldom' in May 1993. In the laboratory, specimens in an early stage of gravidity were selected within the range of 20-50 mg fresh wt, and randomly distributed over two experimental treatments and a control. The animals were kept individually in glass Petri dishes at 15 °C, and were fed weekly replacements of *Alnus glutinosa* leaf disks, contaminated with phenanthrene (0, 20, or 200 $\mu g.g^{-1}$ dry wt). Each concentration was tested in a 15-fold replication. Growth and some life-history characteristics were followed until the next moult after the gravidity period.

4.2. RESULTS AND DISCUSSION

The experiment has lasted for 10 weeks. During this period two animals were lost, as these drowned in their drinking reservoirs. The results for growth, reproduction and moulting are summarized in Table 4. It can be seen that exposure to phenanthrene led to an increased reproductive effort per female, expressed as the total number of eggs and mancas (prejuveniles) that were aborted or given birth. Any resorption of eggs or mancas was not quantified. The data suggest that an increased number of eggs have developed into mancas in the lower phenanthrene treatment, whereas in the higher treatment, as well as in the control, this development was less successfull, and eggs were more frequently aborted. This effect was reflected in the percentage of successfull broods.

Further, the marsupial period was increased (Table 4, Figure 4). If the observed intervals (established for females that were gravid at the start of the

TABLE 4. Life-history characteristics of *Trachelipus rathkei* females as affected by phenanthrene during the marsupial period.

Characteristic	Control	Treatment *(mg FEN.g^{-1} food)*	
		20	200
Total reproductive effort *(female^{-1})*	11.7	13.3	18.2
Live mancas *(%)*	36.9	64.8	29.5
Dead mancas *(%)*	0.6	1.0	1.7
Aborted eggs *(%)*	62.5	34.2	68.8
Successfull broods *(%)*	46.7	60.0	46.2
Marsupial period$^{\#}$ *(days)*	41.7	44.8	47.5
Intermoult$^{\#}$ *(days)*	60.6	57.5	57.2
Biomass at moult *(mg)*	35.5	35.8	36.3
Number of test specimens	15	15	13

$^{\#}$ Note that the experiment was started with pregnant specimens. Experimental values given here. Snider & Shaddy reported 51.4 and 68.3 days for the duration of gravidity and intermoult, respectively (determined at 15.6 °C) [8].

experiment) were corrected to encompass the entire marsupial period as given by Snider and Shaddy [8], the retardation of development rated 6% and 11% in the low and high treatment, respectively. However, the number of days remaining until the next moult was reduced in the treatments, so that all individuals moulted within two weeks, irrespective of treatment. As initial and final biomass did not differ between the experimental concentrations, the effects on reproduction and moulting cannot be ascribed to differences in individual size.

From the weekly data on individual biomass (Figure 4) it is clear that control animals invested more biomass into their brood. Given an increase in number of mancas per female in the low treatment, it is suggested that the individual biomass in the offspring is severely reduced. By consequence, vitality may also be small, as is suggested by higher mortality rates for mancas in the high treatment (Table 4).

5. General discussion

Our preliminary field study reveiled that PAH were taken up by soil animals. Specific variation in total PAH body burden was high. Therefore, for bioindication purposes suitable species or species groups should be selected to monitor environmental risks of soil pollution. Both bioaccumulation and ability for

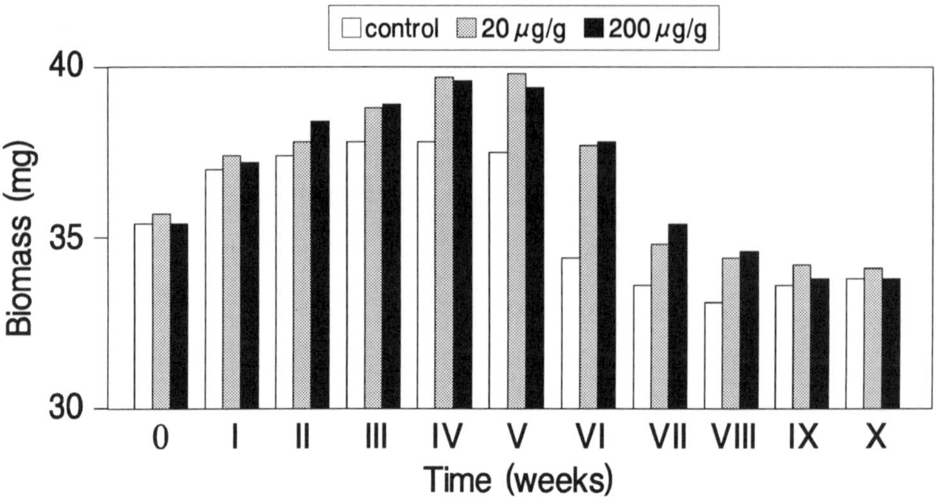

Figure 4. Biomass of female *Trachelipus rathkei* at weekly intervals under exposure to phenanthrene.

biotransformation, with consequent increasing risks for intoxication, must be addressed by the bioindicator system. A feeding guild analysis suggested that wood consumers show a stronger tendency for PAH accumulation than other groups. Within guild differences in concentration factors for individual PAH compounds were also detected. Our laboratory studies confirm earlier reports that isopods are susceptable to PAH (metabolites) intoxication [12, 14]. In view of the hypothesis depicted in Figure 1, isopods can therefore be considered as good representatives for the soil detritivore community.

In conclusion, the use of soil detritivores in bioindication systems promises to be a valuable tool to assess ecotoxicological risks of PAH contamination of soils, both at the community level, as well as at the functional level. This functional group is largely exploited as a food source by higher trophic levels, and is essential in various soil processes.

6. Acknowledgements

Jaap Immerzeel skillfully performed PAH analyses. The identification of enchytraeid species by Wim Didden (Agricultural University, Wageningen) is gratefully acknowledged. Joke van Wensem and Jaap Immerzeel suggested improvements to an earlier version of the manuscript.

7. References

1. Bowmer, C.T., Roza, P., Henzen, L. and Degeling, C. (1993) The development of chronic toxicological tests for PAH contaminated soils using the earthworm *Eisenia fetida* and the springtail *Folsomia candida*. TNO, Delft, The Netherlands, Report nr. IMW-R 92/387.
2. Edwards, N.T. (1983) Polycyclic aromatic hydrocarbons (PAH's) in the terrestrial environment - a review. *J. Environ. Qual.* **12**, 427-441.
3. Kaplan, D.L. and Hartenstein, R. (1978) Studies on the monooxygenases and dioxygenases in soil macroinvertebrates and bacterial isolates from the gut of the terrestrial isopod, *Oniscus asellus* L. *Comp. Biochem. Physiol. B***60**, 47-50.
4. Ministry of Agriculture, Nature Management and Fisheries (1990) *Regeringsbeslissing Natuurbeleidsplan*. Tweede Kamer, Vergaderjaar 1989-1990, nr. 21149, Staatsuitgeverij, Den Haag (in Dutch).
5. Morello A., Bleecker, W. and Agosin, M. (1971) Cytochrome P-450 and hydroxylating activity of microsomal preparations from whole houseflies. *Biochem. J.* **124**, 199-205.
6. Neuhauser, E.F. and Callahan, C.A. (1990) Growth and reproduction of the earthworm *Eisenia fetida* exposed to sublethal concentrations of organic chemicals. *Soil Biol. Biochem.* **22**, 175-179.
7. Pathirana, S., Connell, D.W. and Vowles, P.D. (1994) Distribution of polycyclic aromatic hydrocarbons (PAHs) in an urban roadway system. *Ecotoxicol. Environ. Safety* **28**, 256-269.
8. Snider, R. and Shaddy, J.H. (1980) The ecobiology of *Trachelipus rathkei* (Isopoda). *Pedobiologia* **20**, 394-410.
9. Tebaay, R.H., Welp, G., Brummer, G.W. (1993) Contents of polycyclic aromatic hydrocarbons (PAH) and their distribution pattern in soils with different contamination levels. *Zeitschrift Pflanzenernähr. Bodenk.* **156**, 1-10.
10. Van Straalen, N.M., Verweij, R.A. (1991) Effects of benzo(a)pyrene on food assimiliation and growth efficiency in *Porcellio scaber* (Isopoda). *Bull. Environ. Contam. Toxicol.* **46**, 134-140.
11. Van Straalen, N.M., Verweij, R.A. and Van Brummelen, T.C. (1993) PAH concentrations in forest floor invertebrates in the vicinity of a blast furnace plant. In: Ph. Garrigues and M. Lamotte (eds.) *Polycyclic Aromatic Compounds. Synthesis, Properties, Analytical Measurements, Occurrence and Biological Effects*. Proc. XIIIth Int. Symp. Polynuclear Aromatic Hydrocarbons, Bordeaux, France, Oct. 1-4 1991 Gordon & Breach Sci Publ., pp. 1001-1006.
12. Van Brummelen, T.C. and S.C. Stuijfzand, (1993). Effects of benzo(a)pyrene on survival, growth and energy reserves in the terrestrial isopods *Oniscus asellus* and *Porcellio scaber*. *Sci. Total Environ.* **Suppl.**(1993), 921-930.
13. Van Brummelen, T.C., Verweij, R.A., Van Straalen, N.M. (1991) Determination of benzo(a)pyrene in isopods (*Porcellio scaber* Latr.) exposed to contaminated food. *Comp. Biochem. Physiol. C***100**, 21-24.
14. Van Hattum, B. (1995) *Bioaccumulation of sediment-bound contaminants by the freshwater isopod* Asellus aquaticus *(L.)*. Thesis Vrije Universiteit, Amsterdam, The Netherlands.
15. Wilbrink, M. (1991) *The pond snail* Lymnaea stagnalis *in toxicological research. Fate and effects of dihalogenated biphenyls*. Thesis Vrije Universiteit, Amsterdam, The Netherlands.

SUBLETHAL TOXICITY TESTS FOR LONG-LIVED ITEROPAROUS INVERTEBRATES: SEARCHING FOR A SOLUTION

R. LASKOWSKI[1], M. MARYAŃSKI[1],
E. PYZA[2] and J. WOJTUSIAK[2]

[1]*Department of Ecosystem Studies*
Institute of Environmental Biology
[2]*Zoological Museum, Institute of Zoology*
Jagiellonian University
Ingardena 6, 30-060 Kraków, Poland

1. Introduction

Many invertebrates of major importance for the functioning of terrestrial ecosystems are long-lived iteroparous species. They range from herbivores (e.g., snails and slugs) through carnivores (e.g., centipedes) to detritivores (e.g., earthworms, diplopods). From an ecotoxicological point of view, all these groups deserve special attention for at least two reasons: their well documented importance for ecosystem function [1, 2, 3, 4], and the fact that many of them can accumulate substantial amounts of pollutants in their tissues [5, 6, 7, 8, 9, 10]. The latter point may have two important consequences: these animals can be more endangered by pollution than other, short-lived organisms, and they may constitute a critical pathway for pollutants along trophic chains (*cf.* [11]).

Environmental pollution can be a strong stressing factor for soil organisms. Stress is understood here as defined by Sibly and Calow [12]: an environmental condition that reduces Darwinian fitness. Thus, the general problem of ecotoxicology is to estimate pollution effects on average fitness in populations. Unfortunately, the whole life cycle of long-lived iteroparous invertebrates takes several years and it is impossible to construct their life-tables during standard short laboratory experiments. One of the solutions proposed so far employs methods based on so-called biomarkers, for example heat-shock proteins. These methods, although potentially useful in detecting whether a particular organism is under stress, do not offer much insight into the actual effects of toxicants at the level of populations or even single specimens. Another method of estimating population-level effects was proposed by Laskowski [13] and is based on life-history studies and a life-table approach. The method permits inferences about population effects for a number of iteroparous species such as earthworms, snails or slugs after relatively short experimental studies lasting for a few weeks up to a few

N. M. van Straalen and D. A. Krivolutsky (eds.), Bioindicator Systems for Soil Pollution, 45–53.

months, but there are still many species for which much longer research would be required. These are species with low reproductive output, without a single breeding or egg-laying season (e.g., centipedes). Due to this specificity, observation of small, sublethal effects of toxicants on life-history parameters would require studies lasting a few years. This is usually unacceptable for standard ecotoxicological tests. Thus the need to search for more practical tools to assess the effects of toxicants on the condition of a specimen and, in longer perspective, on population parameters.

Surviving in a polluted environment is usually connected with changes in metabolic processes. These may result from direct poisoning that blocks and damages enzyme proteins [14, 15] or from employment of specialized physiological detoxification pathways (*cf.* [8]). Both require synthesis of proteins, which is believed to be an energetically costly process [16, 17]. Thus metabolic changes may serve as physiological biomarkers of stress in intoxicated organisms.

Migula and other authors have shown that enzyme activity and respiratory metabolism can be used to detect the effects of pollution at both tissue and specimen levels [18, 19, 20, 21]. Unlike biochemical biomarkers such as metallothioneins or heat-shock proteins, these physiological indices may be easier to relate to the physiological condition of a specimen and may permit estimation of the energetic costs of living in a polluted environment. Thus they seem especially useful for ecotoxicological studies of long-lived iteroparous species for which other, more direct methods are not available.

In this paper we discuss the results of studies on the respiration rate and locomotor activity pattern and level in a carnivorous centipede (*Lithobius mutabilis*) as possible tools for assessing sublethal effects of toxicants on long-lived iteroparous invertebrates.

2. Methods

Populations of field-captured centipedes (*L. mutabilis* Koch., Chilopoda) were maintained in the laboratory at 15°C±0.2°C for a few months. Adult centipedes of 15-45 mg fresh mass (10-15 mm length) were used in the experiment. Single animals were kept in 11 × 7.5 × 4.5 cm plastic containers, and for population studies two males and two females were kept in 16 × 11 × 6 cm boxes. The boxes had a 1-cm layer of standard artificial soil (OECD; moisture = 50% water holding capacity) and were covered with loose transparent plastic lids. The animals were kept under a 16:8 light:dark regime.

Two toxicants were used in the experiment: copper (Cu) as an example of a persistent pollutant of industrial origin, and dimethoate (DMT) as an example of a degradable insecticide.

In the experiments with Cu, the centipedes were kept on contaminated soil and fed twice a week with a half of a frozen woodlouse (*Porcellio scaber*) per individual. To make contaminated food for the centipedes, the woodlice were fed Cu-contaminated artificial food (oak-hornbeam leaf litter + rabbit chow + water). The nominal concentrations of Cu in the soil and food offered to *P. scaber* were 0 (control), 10, 40,

160 and 640 mg kg^{-1} dry weight of soil/food. Four replicates per treatment were used in both single-specimen and population studies.

In the experiments with DMT, housefly pupae were used as food. Two runs of the experiment were done: one on the effects of contaminated food on centipedes (DMT-I) and the other on the effects of contaminated soil (DMT-II). For DMT-I, housefly sensitivity to DMT determined the upper limit of concentrations used: 0.8 mg a.i. kg^{-1} dry mass of food. Thus the concentration used in DMT-I were 0 (control), 0.037, 0.11, 0.33 and 0.8 mg a.i. kg^{-1}. As no effects on centipedes were found in this experiment, for DMT-II higher concentrations were used: 0, 1.56, 3.1, 6.25, 12.5 and 25 mg a.i. kg^{-1} dry soil. In the DMT-I and DMT-II experiments six replicates were used per treatment.

The variables measured were mortality, consumption rate (only for Cu), growth rate, respiration rate, locomotor activity level and pattern, and Cu concentration in *P. scaber* and *L. mutabilis* (atomic absorption spectrometer; Cu only). Only the results on respiration rate and locomotor activity are discussed here. The experiment was terminated after 127 days for Cu and after 30 days for DMT. The respiration rates of Cu-treated centipedes were measured 8, 26, 43, 72 and 99 days after the beginning of the experiment, and weekly for DMT-treated ones. Manometric microrespirometers in a water bath at 15°C±0.2°C were used. Activity level and pattern were measured at the termination of the experiment with digital image processing methods (Oxalis software).

3. Results

Isopods fed a Cu-contaminated diet and used as prey for centipedes showed a linear increase in total body concentrations of Cu with increasing Cu concentration in food (Fig. 1). Copper concentrations in centipedes fed Cu-contaminated isopods also had a clear increase compared to the control, but not a linear one. There was a significant increase in Cu concentration at the first treatment, and approximately the same elevated level of Cu was maintained in the animals from all the experimental treatments at and above 10 mg kg^{-1}, that is, at nominal concentrations in the soil and food provided to isopods 10 - 640 mg kg^{-1} (Fig. 1).

A significant decrease (regression analysis) in the respiration rate of Cu-treated centipedes was found on the 26th day after the start of the experiment ($p<0.0005$) (Fig. 2). Later the respiration rates of the treated animals returned to the level of those in the control treatments. No effects of DMT-contaminated food were found in the DMT-I experiment. In DMT-II, significant differences (ANOVA) in respiration rates between experimental groups were found 1, 8, 15 and 22 days after the start of the experiment (Fig. 3), but no significant difference was detected after 29 days, nor for the average respiration rate for the whole experimental period. Even when the differences were

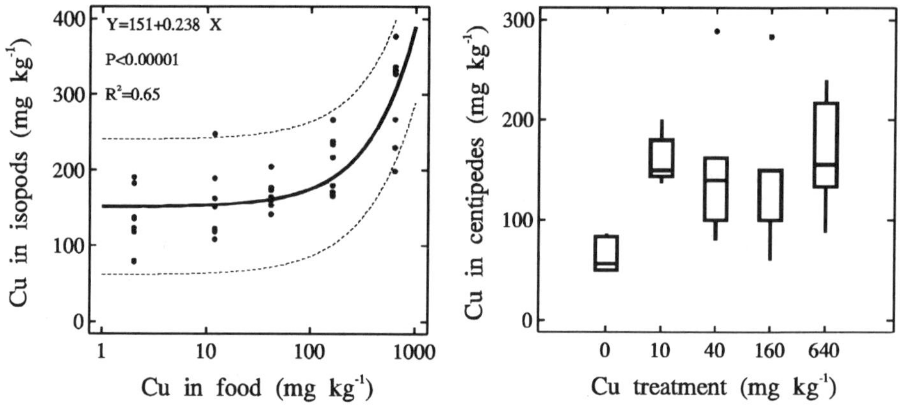

Figure 1. Copper concentrations in isopods and centipedes exposed to Cu-contaminated diet and soil; Cu treatment - the nominal Cu concentrations in the soil and the diet of prey isopods.

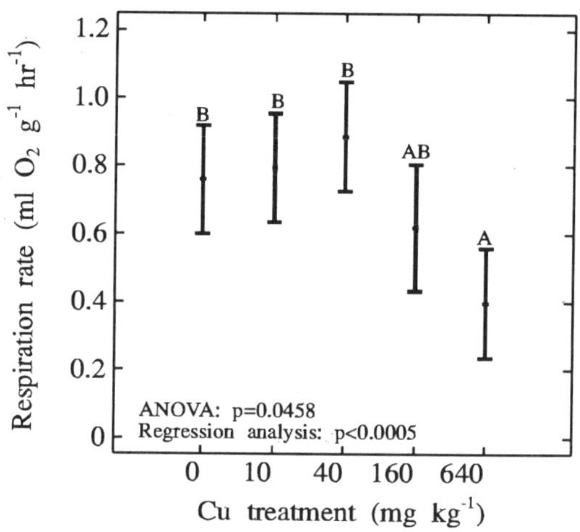

Figure 2. Respiration rate of centipedes vs. Cu treatment (95% LSD intervals); Cu treatment - nominal concentrations in the soil and in the diet of prey animals.

significant, no clear relation between respiration rate and soil DMT concentration could be found; regression analysis gave non-significant results for all respiration measurements. The centipedes treated with the highest concentration of DMT (25 mg a.i. kg^{-1} soil) had the lowest respiration rate among all treatments at days 2, 16 and 23, but at day 9 they had exactly the same respiration rate as the control animals (Fig. 3).

Regression analysis revealed a significant decrease in average daily locomotor activity with increasing Cu concentration in centipedes ($p<0.05$, $R^2=0.78$). After excluding two outliers (one exceptionally low value for the control treatment and one exceptionally high one for 640 mg kg^{-1}), the significance level decreased to $p<0.027$

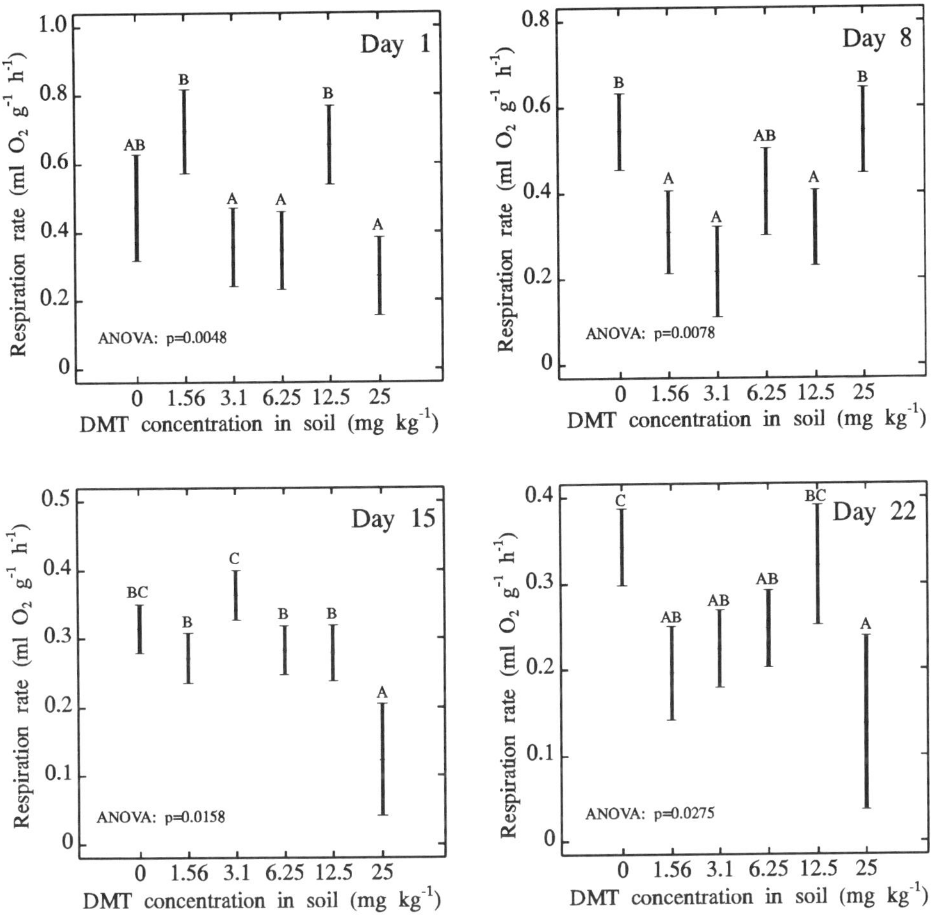

Figure 3. Respiration rate of centipedes vs. nominal DMT concentration in soil (mg a.i. per kg dry weight) 1, 8, 15 and 22 days after the start of the DMT treatment (95% LSD intervals).

(R^2=0.84). However, both regressions were driven mainly by one point, with a high level of activity for the control animals (35% active time daily or 48% after excluding the outlier) and with all the other treatments heaving similar activity levels (19%-25% active time). Exclusion of the outlier from the highest treatment gave the lowest activity level, 9% daily (Fig. 4). In the DMT-II experiment no differences were found between treatments (Tab. 1).

The pattern of locomotor activity also seems to be affected by the highest Cu treatments. While the control animals revealed a relatively well pronounced day/night rhythm and activity pattern, with both day-active and night-active individuals, no clear pattern or rhythm could be found in animals from the highest treatment. Those centipedes were not able to synchronize their activity to changes in light conditions under the day/night cycle. No visible changes in the activity pattern were noted in the

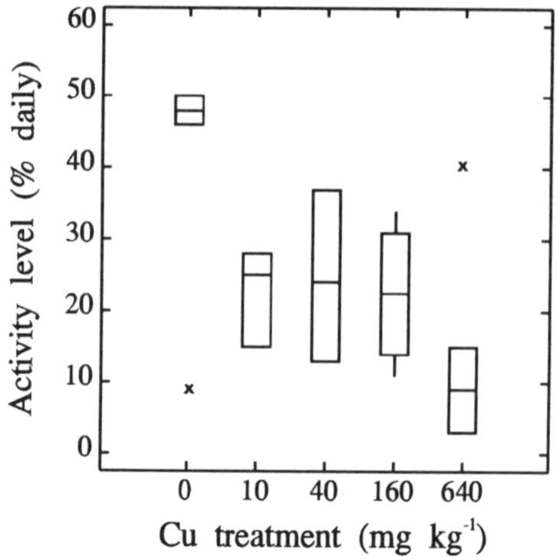

Figure 4. Locomotor activity level in centipedes vs. Cu treatment (average percent active time daily). Two excluded outliers are marked with ×; see text for details.

DMT-treated animals.

TABLE 1. Locomotor activity level in cetipedes at different dimethoate (DMT) treatments (average percent active time daily during one-week measurements ± SD).

DMT concentration in soil mg a.i. kg^{-1}	Activity level average±SD (% daily)
0 (control)	19.4±15.9
0.012	27.5±10.3
0.037	26.6±20.1
0.111	15.2±9.8
0.333	27.0±12.5
0.800	21.3±15.2

4. Discussion

The results on Cu and DMT toxicity to centipedes indicated that measurements of respiration rates and locomotor activity levels and patterns may be promising tools for monitoring sublethal effects of pollution in long-lived invertebrates. Decreased respiration rates in semelparous insects exposed to various kinds of environmental pollutants has been shown by [18, 22, 23]. In our studies, however, the respiration rate of centipedes treated with Cu and DMT was affected for only a relatively short period. Additionally, the pattern of the relation between the respiration rate and soil concentration of DMT changed considerably over time. Thus the measurements have to be carried out at intervals not longer than those used in our study, at least approximately every 20 days for persistent chemicals such as heavy metals and at least weekly starting from the first day after treatment for degradable chemicals such as insecticides.

There are two different possible responses to intoxication in the respiration rate: a decrease due to damage in respiratory enzymes and an increase due to the increased costs of surviving in a polluted environment. Theoretically a single animal can go through both of them, especially in the case of slowly accumulating persistent chemicals. First, during the accumulation phase the percentage of damaged enzymes should rise, decreasing the respiration rate at this stage. Next, when physiological mechanisms for detoxification are activated, metabolism should return to its initial level and eventually increase above the normal level. If the tolerance threshold is exceeded, the respiration rate decreases again and the animal dies. In our studies on Cu we observed only the first stage, the decrease in the respiration rate, and a return to the normal metabolic level. Longer studies would be required to investigate the third phase of an increased metabolic rate in intoxicated animals. The results of the DMT-II experiment are more difficult to interpret as no clear relation between concentration and effect was found, but Figure 3 suggests that all four phases were detected for the highest DMT treatment (25 mg a.i. kg^{-1} soil), without metabolism exceeding the normal level. The respiration rate decreased 24 hours after the start of the exposure, returned to the normal (control) level after the next week, decreased again at day 16 and remained significantly below the control level at day 23, with only one centipede alive. No centipede treated with 25 mg a.i. kg^{-1} soil survived to the last measurement, 29 days after the start of the experiment.

The effects of sublethal doses of pollutants on the locomotor activity level and pattern are practically unknown. Our studies indicate that the main problem with using them as indices of pollution stress is their large natural variability in both unaffected and affected populations (see Fig. 4 and Tab. 1). With a large number of replicates and a carefully planned measurement schedule, however, they may be useful tools for assessing sublethal effects, at least for some groups of invertebrates. For persistent chemicals such as heavy metals, the experimental design used in our study, with activity measurements at the termination of treatment, seems to work, but for fast-acting toxicants such as insecticides the activity observations should be made continuously rather than at the end of the exposure period.

5. Conclusions

1. Significant effects on both respiration rates and locomotor activity levels and patterns were found in centipedes exposed to toxicants.
2. Respiration rates and locomotor activity levels and patterns appeared highly variable in untreated and treated animals. Additionally, the effects of pollutants on the respiration rate, and probably also on locomotor activity, changed over time. To investigate the sublethal effects of low concentrations of pollutants, the measurements mast be made

frequently (at 1-3-week intervals, depending on the chemical) on a large number of replicates.
3. Among the two methods tested, the respiration rate measurements seemed easier to perform and less variable than were the locomotor activity level and pattern measurements.
4. To make any inferences about the population effects of toxicants using data on respiration rates and locomotor activity patterns and levels, these data must be linked to experiments on mortality and fertility rates. In animals with long and complicated life histories, such as centipedes, some indirect methods for estimating pollution effects on fertility have to be developed. Preliminary studies suggest that one possibility could be observations of pollution effects on ovaries.

Acknowledgements

The funding for this work was provided by the EU Programme Environment 1990-1994 (the PECO Programme contract no. ERB-CIPD-CT93-0059).

6. References

1. Albert, A.M. (1976) Biomasse von Chilopoden in einem Buchenaltbestand des Solling, *Verhandlungen der Gesellschaft für Ökologie*, Göttingen: 93-101.
2. Jennings, T.J. and Barkham, J.P. (1979) Litter decomposition by slugs in mixed deciduous woodland, *Holarctic Ecology* **2**, 21-29.
3. Albert, A.M. (1983a) Energy budgets for populations of long-lived arthropod predators (Chilopoda: Lithobiidae) in an old beech forest, *Oecologia (Berl.)* **56**, 292-305.
4. Rożen, A. (1989) The annual cycle in populations of earthworms (Lumbricidae, Oligochaeta) in three types of oak-hornbeam of the Niepołomicka Forest. II. Dynamics of population numbers, biomass and age structure, *Pedobiologia* **31**, 169-178.
5. Hopkin, S.P. and Martin, M.H. (1983) Heavy metals in the centipede *Lithobius variegatus* (Chilopoda), *Environ. Pollut. (B)* **6**, 309-318.
6. Bengtsson, G., Gunnarsson, T. and Rundgren, S. (1986). Effects of metal pollution on the earthworm *Dendrobaena rubida* (Sav.) in acidified soils, *Water, Air, Soil Pollut.* **28**, 361-383.
7. Coughtrey, P.J. and Martin, M.H. (1977) The uptake of lead, zinc, cadmium and copper by the pulmonate mollusc, *Helix aspersa* Müller, and its relevance to the monitoring of heavy metal contamination of the environment, *Oecologia (Berl.)* **27**, 65-74.
8. Hopkin, S.P. (1989) *Ecophysiology of metals in terrestrial invertebrates*, Elsevier Applied Science Publishers Ltd, London.
9. Jones, D. (1991) *Biological monitoring of metal pollution in terrestrial ecosystems*, Ph.D. Thesis, Department of Pure and Applied Zoology, University of Reading, U.K.
10. Laskowski, R. and Maryański, M. (1993) Heavy metals in epigeic fauna: trophic-level and physiological hypotheses, *Bull. Environ. Contam. Toxicol.* **50**, 232-240.
11. Van Straalen, N.M. and Ernst, W.H.O. (1991) Metal biomagnification may endanger species in critical pathways, *Oikos* **62**, 255-256.
12. Sibly, R.M. and Calow, P. (1989) A life-cycle theory of responses to stress, *Biol. J. Linn. Soc.* **37**, 101-116.

13. Laskowski, R. (1995) Estimating fitness costs of pollution in iteroparous invertebrates, in: N. van Straalen and H. Lökke (eds.), *Ecological Principles for Risk Assessment of Contaminants in Soil*, Chapmann and Hall, in press.
14. Webb, M. (1979) *The chemistry, biochemistry and biology of cadmium*, Elsevier, Amsterdam.
15. Pool, M. (1981) Exposure and health effects of cadmium. Part 3. Effects of cadmium on enzyme activities, *Tox. Environ. Chem. Rev.* **4**, 179-203.
16. Meier, P.R., Peterson R.G., Bonds D.R., Meschia G. and Battalgia F.C. (1981) Rates of protein synthesis and turnover in fetal life, *American J. Physiol.* **240**, 320-324.
17. Hawkins, A.J.S. (1985) Relationships between the synthesis and breakdown of protein, dietary absorption and turnovers of nitrogen and carbon in the blue mussel, *Mytillus edilus* (Bivalvia: Mollusca), *Oecologia* **66**, 42-49.
18. Pytasz, M., Migula, P. and Krawczuk, A. (1980) Wpływ emisji przemysłowych na metabolizm ogólny wybranych gatunków zwierząt z rejonu huty "Katowice", *Acta Biologica* **8**, 81-103.
19. Migula, P. (1989a) Bioenergetic indices as indicators of environmental contamination in insects, *Nutri. Ecol. Ins. & Env.* **1989**, 157-166.
20. Migula, P. (1989b) Combined and separate effects of cadmium, lead and zinc on respiratory metabolism during the last larval stage of the house cricket, *Acheta domesticus*, *Biológia (Bratislava)* **44**, 513-521.
21. Migula, P. (1990) The effect of cadmium on mitochondrial respiration of the house cricket (*Acheta domesticus* L.), *Acta Biologica Silesiana* **15**, 95-106.
22. Migula, P. (1985) Wrażliwość wybranych gatunków owadów na skażenia powietrza gazami i pyłami przemysłowymi oraz tolerancja przez nie termicznych zmian środowiska, *Prace naukowe Uniwersytetu Śląskiego w Katowicach* **175**, 115 pp.
23. Byrdy, S., Hurny, J., Migula, P., Wawrzyniak, A. and Bińkowska, K. (1988) Metabolizm tlenowy pszczoły miodnej (*Apis mellifera* L.) zatruwanej deltametryną, *Materiały XXVII sesji naukowej Instytutu Ochrony Roślin* **1988**, 87-91

POSSIBLE APPLICATION OF FUZZY SYSTEM SIMULATION MODELS FOR BIOMONITORING SOIL POLLUTION IN URBAN AREAS

G.P. STAMOU* and G.B. STAMOU**
**Department of Ecology*
Aristotle University of Thessaloniki
PUB 119, 540 06 Thessaloniki, Greece
***Intelligent, Robotics and Control Unit*
National Technical University of Athens
15 773, Zografou campus, Athens, Greece

Abstract

Life-history traits of soil arthropods are considered useful for bioindication purposes in urban sites. To model the pollution in these sites on the basis of life-history parameters, highly complex and over-mathematized models are needed. As an alternative, fuzzy system simulation techniques can be applied. In this paper a fuzzy theory-based model, involving temperature and pollution levels in a Mediterranean city is developed. The model was built upon hypotheses drawn from laboratory evidence. Oscillating temperature regimes and pollution levels are the driving forces behind the responses of life-history parameters and determine the magnitude of these responses. A qualitative lay-out of the simulation model is presented.

1. Introduction

Plants and arthropod species can be used as biomonitors of urban areas. According to Denneman and Van Straalen [1] there are indications that some soil invertebrates and especially mite species are sensitive to toxic substances. Moss-dwelling species have been already used for assessing and even for monitoring air pollution [2, 3, 4].

The life-history traits of a species determine the potential range of biotopes that the species can inhabit. Siepel [5] illustrated the practical applicability of defined life-history tactics in ecotoxicology by evaluating the effect of pollution on the soil microarthropod community, on the basis of life-history tactics of the species. Moreover, Joosse and Verhoef [6]

N. M. van Straalen and D. A. Krivolutsky (eds.), Bioindicator Systems for Soil Pollution, 55–65.

and Kooijman and Metz [7] considered the reproductive output as an important parameter in ecotoxicologial research because of its influence on population dynamics. Finally, Schaefer [8] argued that the ecophysiological and population features of the species (*e.g.* tolerance, colonisation ability, length of life-cycle, r or K strategy characteristics) are the main determinants of presence and abundance of species in urban soils.

Stamou & Argyropoulou [9] tried to apply these ideas in a study on the effects of pollution of urban soils with Cu, Pb, and Zn on oribatid mites along a traffic gradient of the city of Thessaloniki. They actually showed that morphometric characteristics as well as certain life-history traits were important for understanding the pollution impact on life in soils. Nevertheless, in spite of the amount of information on the relationship of pollution with life-history tactics, to our knowledge, efforts to model the effects of pollution on population dynamics of arthropods are completely missing. In this paper we examine the possibility to model population dynamics of arthropods inhabiting urban areas by using fuzzy system simulation models.

2. Modelling population dynamics

A Leslie-type model was successfully used to simulate the population dynamics of an oribatid mite from a deciduous Greek forest [10]. Apart from its simulation capacities, the model also involved the demographic parameters fecundity, *f*, mortality, *m*, and life-stage duration, *l*, so that it captures all the information on the animal's life-history tactics. Leslie-type models describe the dynamics of structured populations as follows:

$$\mathbf{N}_{t+1} = \mathbf{A}\ \mathbf{N}_t$$

where N_{t+1} and N_t are demographic vectors describing the population structure at time $t+1$ and t, respectively, and $\boldsymbol{A}$ is the transition matrix. The demographic vectors take the form:

$$\mathbf{N}_t = \begin{bmatrix} N_1 \\ N_2 \\ - \\ - \\ N_i \\ - \\ - \end{bmatrix}$$

where N_i stands for the population size of life-stage i (egg, larva, *etc.*) at time t, while the transition matrix has the form:

$$\mathbf{A} = \begin{bmatrix} S_1 & 0 & 0 & -- & f_i \\ P_1 & S_2 & 0 & 0 & 0 \\ 0 & P_2 & S_3 & 0 & 0 \\ -- & -- & -- & -- & -- \\ 0 & 0 & 0 & P_{i-1} & S_i \end{bmatrix}$$

where f_i is the fecundity of age class i. The diagonal elements S_i represent the probability of an individual not to develop into the next life-stage within the time period $\{t, t+1\}$. The subdiagonal elements P_i account for the probability of a specimen in life-stage $i-1$, at time t, to hatch into the next life-stage during $\{t, t+1\}$. The diagonal and subdiagonal elements of the transition matrix are functions of the duration, l_i, and the mortality rate m_i:

$$S_i = (1-m_i)\frac{l_i-1}{l_i} \quad \text{and} \quad P_i = \frac{1-m_i}{l_i}$$

3. The effect of pollution and temperature on demographic parameters

In Greek Mediterranean areas the organisation of arthropod life-cycles is synchronised with seasonally oscillating climatic variables, especially temperature. Thus, the effect of contamination on soil animals is expected to be time- and temperature dependent. To describe the effect of changing temperatures on the demographic parameters (D_p) of soil arthropods a four parameter model was used:

$$D_p = a\{(1 + k\, e^{-pT})^{-1} - e^{-(T_1-T)}\}$$

where T = temperature, T_l = lethal temperature, and k and p are parameters defining the optimal temperature range (plateau), as well as the inflection point of the curve [11, 12, 13]. Pollution is expected to affect not only the lethal temperature but also to shorten the optimal range as well as to shift the inflection point of the curve (Fig. 1).

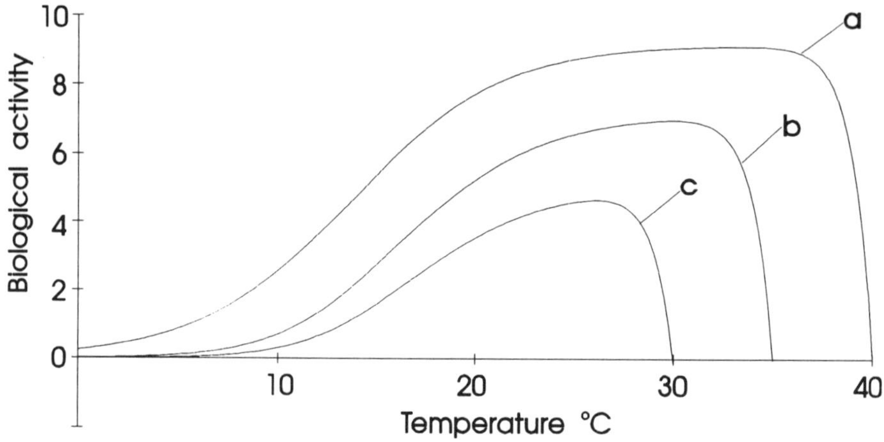

Figure 1. The effect of temperature on the biological activity of soil arthropods in unpolluted (a), polluted (b) and highly polluted (c) areas.

Experimental data on the oribatid mite *Scheloribates cf. latipes* [14] have shown that contaminating the substrate and the food affects both the relationship of survivorship with age and the relationship of fecundity with age. Survivorship data (S, in %) were described by means of a two parameter model:

$$S = 100(1-t^{a})^{b/t}$$

where t = time and a and b are parameters accounting for the effects of contamination on survivorship. By changing the parameters a family of curves can be obtained (Fig. 2). From Fig. 2 it can be seen that the contamination of the substrate and food mainly affects mortality of younger specimens, shortens the animal's life-span and changes the shape of the survivorship curve.

The effects of substrate and food contamination on fecundity were described using the following six parameter empirical model:

$$F = a_0\{(1 + a_1 e^{-b_1 t})^{-b_2} - c_0\, e^{c_1 - t}\} - d$$

Again, a family of curves is obtained by changing the parameters. Observations showed that contamination shortened the oviposition period, delayed the onset of the oviposition and lowered the fecundity rate (Fig. 3).

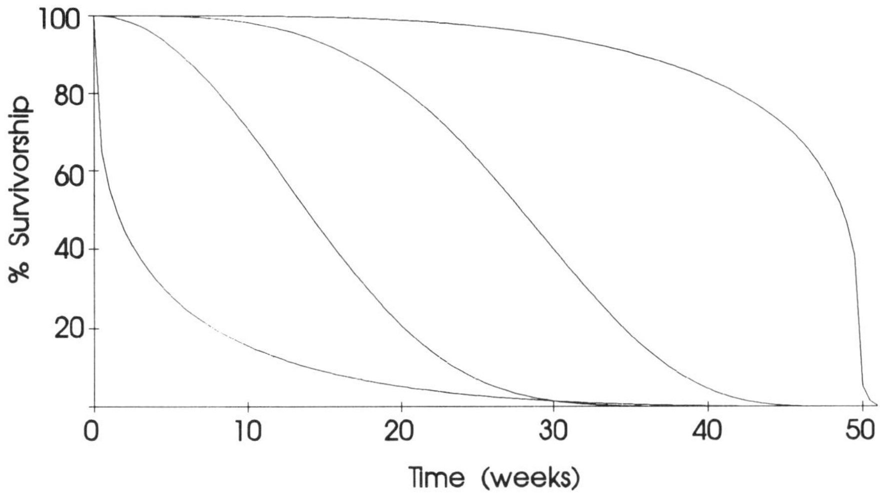

Figure 2. The effect of contamination on the relationship of survival with age in cultures of oribatid mites. In clean cultures, concave curves were obtained, while in highly contaminated cultures a convex curve was seen. Cultures at intermediate levels of contamination produced sigmoid graphs.

From the above given considerations it becomes obvious that pollution acts either directly or indirectly through temperature, on demographic parameters and affects the type as well as the magnitude of the response. Thus pollution is the major selective force, while temperature is the force driving the expression of the response.

4. Fuzzy model building

Urban areas form a mosaic of more or less isolated biotopes with different levels of pollution in which a variety of species can be found. To assess the effect of soil contamination on the demographic traits of soil arthropods using the model explained above, twelve parameters (four relating to temperature, two to survivorship and six to fecundity) should be taken into consideration for each species at each site. Due to the high diversity of community structures in connection with the high diversity of the biological processes involved it appears difficult to simulate the dynamics of real systems by using conventional mathematical tools. Thus any assessment of the effects of soil pollution appears highly complex.

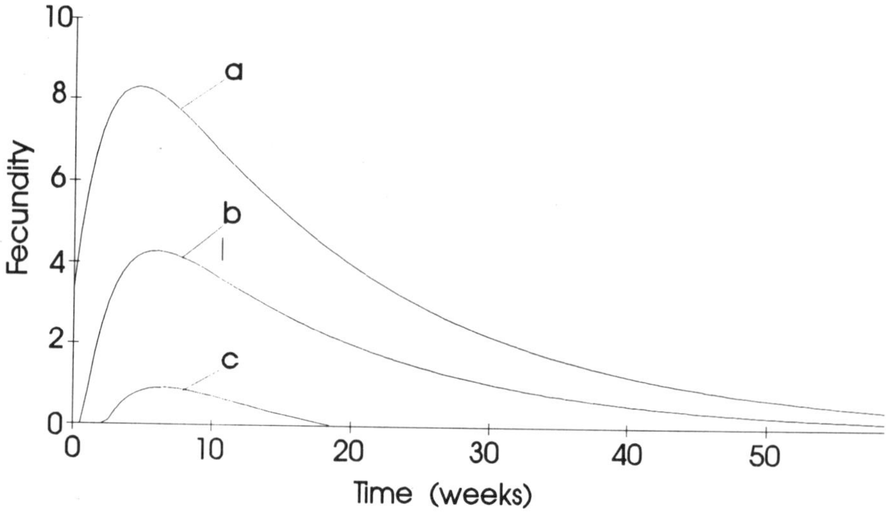

Figure 3. The relationship of fecundity with time. a: low pollution level, b: intermediate pollution level, c: high pollution level.

According to Levins [15] ecological modelling has to find a compromise between biological realism, generality and precision. To obtain a realistic and general, although approximate, solution of the pollution problem, fuzzy systems simulation techniques, which lower the complexity of over-mathematized situations, appear promising. Fuzzy systems were developed to tackle problems arising within the artificial knowledge framework; the theory aims to simulate the mechanisms of human knowledge. The great advantage of fuzzy system theory in ecology is due to the fact that it replaces arithmetic variables with linguistic ones, such as: "very low pollution", "intermediate pollution", "highly polluted", *etc.*

Fuzzy theory claims that words, especially adjectives, are ambiguous with respect to their exact meaning [16]. Consider, for example, the case in which for a given biotope the instruments record a temperature of 20 °C. In most microsites the actual temperature will be 20 °C indeed, while in some other microsites the actual temperature can be 25 °C or 15 °C. Thus the actual temperature in the different microsites within the biotope is distributed around 20 °C and it can be characterised as "medium", which it means that it is around 20 °C, although it might be also 25 °C or 15 °C. As a consequence, in reality there is an amount of ambiguity connected to the recorded temperature of 20 °C.

To cope with this ambiguity, the function of a fuzzy system is based upon the definition of the membership function $\mu_A(x)$ where X is the

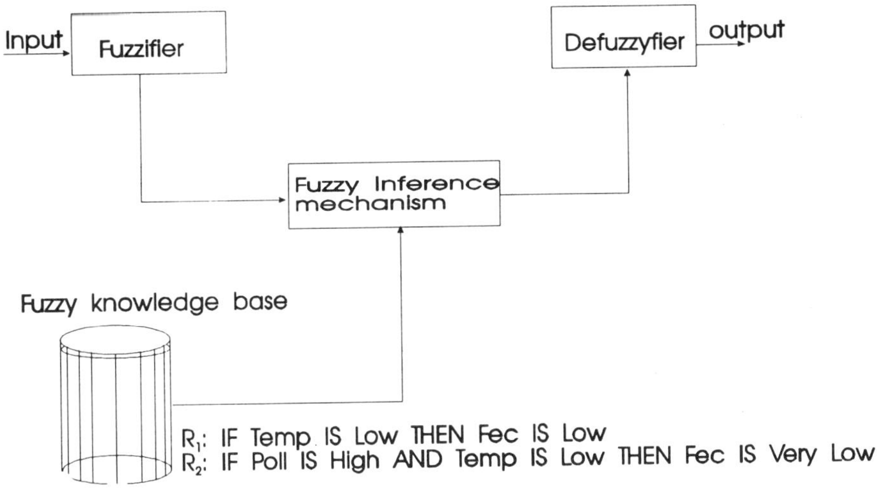

Figure 4. The structure of the fuzzy system. The fuzzyfier, the knowledge base, the fuzzy inference mechanism and the defuzzifier are the compartments of the system. Two rules contained in the fuzzy knowledge base are given by way of example.

reference set, called "universe of discourse" (*e.g.* the temperature gradient: *very low, low, relatively low, intermediate, relatively high, high, very high temperatures*), *A* is a subset of *X* (*e.g. high temperature*) and *x* is a given object (*e.g.* 20 °C). The membership function $\mu_A(x)$ assigns to an object *x* (*e.g.* the given temperature of 20 °C) a weight that measures the probability that this object belongs to the subset *A* (*e.g.* high temperature). The subset *A* is formulated as follows:

$$A = \{(x, \mu(x)): \; x \in X\}$$

in the case of a continuous universe of discourse, while in the case of a discrete universe of discourse the subset *A* has the form:

$$A = \mu(x_1)/x_1 + \mu(x_2)/x_2 + ... + \mu(x_n)/x_n$$

For example, the subset *A* = *high temperature* may be formulated as:

$$A(\text{high temperature}) = 0/0°C + 0/10°C + 0.3/20°C + 0.6/30°C + 1/40°C.$$

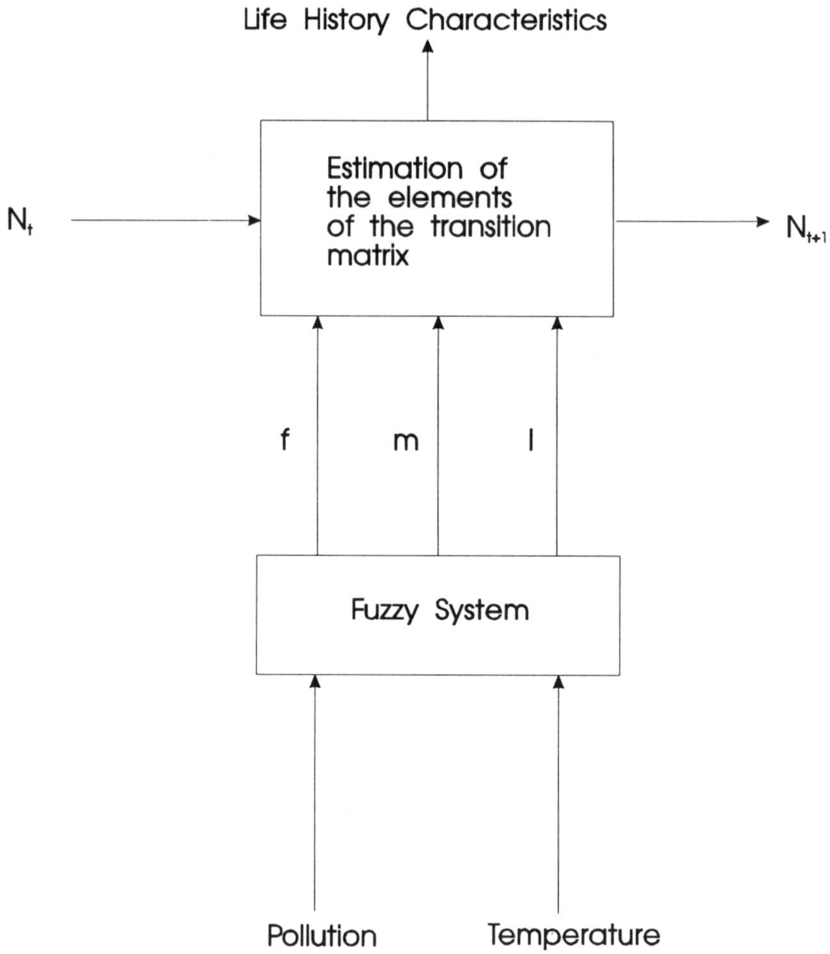

Figure 5. Lay-out of the simulation model for the study of effects of temperature and pollution on the dynamics of arthropod populations in urban areas.

Any fuzzy system is made up of four units, the "Fuzzifier", the "Fuzzy Knowledge Base", the "Fuzzy Inference Mechanism" and the "Defuzzifier" (Fig. 4). Fuzzy inference is based upon simple rules contained in the Fuzzy Knowledge Base. For example, the following sentence:

if temperature **is** low **and** pollution **is** high **then** mortality **is** medium

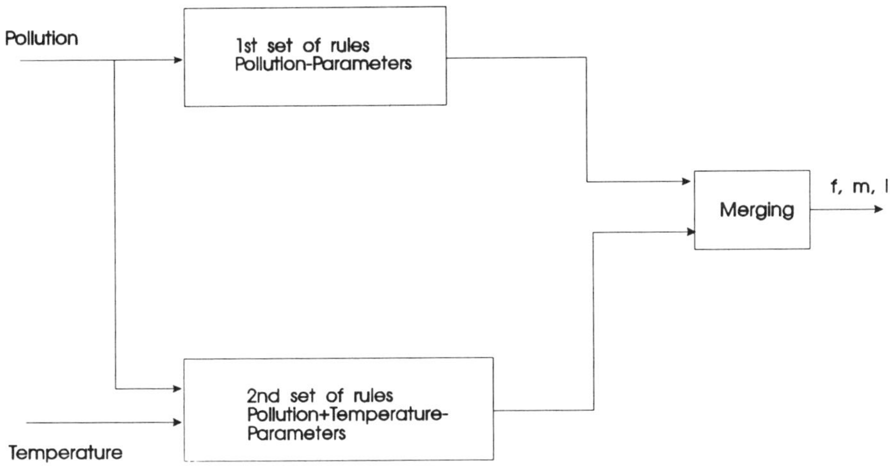

Figure 6. The structure of the fuzzy subsystem of the simulation model.

is such a rule. In the fuzzification unit the arithmetic variables which constitute the system inputs are transformed into linguistic variables, while the fuzzy inference mechanism provides output by using the generalised *modus ponens* rule: $(A \rightarrow B \wedge A') \rightarrow B'$. This rule says that if object A involves object B and if an object A', comparable to A, occurs then this will involve object B' which is comparable to B. Finally, in the defuzzification compartment the linguistic variables can be transformed back into arithmetic ones.

The structure of the actual simulation model is depicted in Fig. 5. Observations on pollution and temperature are the inputs to the fuzzy compartment and the demographic parameters are the outputs. In turn, the latter are the inputs to the compartment in which the compilation of the transition matrix takes place. Then the demographic vector $\boldsymbol{N}_t$ is multiplied by the transition matrix $\boldsymbol{A}$ to produce the demographic vector $\boldsymbol{N}_{t+1}$.

The structure of the fuzzy subsystem is given in Fig. 6. The direct effects of pollution on demographic parameters are specified in the first set of rules. The indirect effects on demographic parameters (through temperature) are specified in the second set of rules. The outputs are then merged to produce the actual linguistic values for the demographic parameters (fertility, mortality, and stage duration).

Apart from the demographic vector $\boldsymbol{N}_{t+1}$ the model also produces a table of life-history characteristics. Thus, an additional simulation output could be, for example, as follows: "*Populations with a high rate of development into next stages, a peak in fecundity, early reproduction and high mortality among adults are expected to be sampled in June*". When a monthly sequence of life-history characteristics is made up, a time-dependent life-history schedule for arthropods in urban environments could be obtained.

The conclusions which can be drawn from the above presentation are the following:

1. The inherent complexity of over-mathematized systems will be avoided, since the most difficult part, the dynamics of demographic parameters, is described qualitatively using linguistic variables and rules.

2. This approach gives more realistic descriptions of ecological systems. For example it could describe the case in which a recorded temperature of 20 °C is actually 20 in some microsites, while in others it 15 or 25 °C.

3. The model can be easily adapted to different organisms and biotopes by simply modifying the rules and the initial values of the demographic parameters and demographic vectors.

5. References

1. Denneman, C.A.J. and Van Straalen, N.M. (1991) The toxicity of lead and copper in reproduction tests using the oribatid mite *Platynothrus peltifer*, *Pedobiologia* **35**, 305-311.
2. Lebrun, Ph., Wauthy, G., Leblanc, Ch. and Goossens, M. (1977) Tests écologiques de toxitolerance au SO_2 sur l'oribate corticole *Humerobates rostrolamellatus* GRANDJEAN 1936 (Acari: Oribatei), *Annales Soc. r. Zool. Belg.* **106**, 193-200.
3. Lebrun, Ph., Jacques, J.M., Goossens, M. and Wauthy, G. (1978) The effects of interaction between the concentration of SO_2 and the relative humidity on the survival of the bark-living bioindicator mite *Humerobates rostrolamellatus, Water Air Soil Pollut.* **10**, 269-275.
4. André, H.M., De Sloover, J., Iserentaut, R. and Lebrun, Ph. (1984) The use of biological monitors of acid air pollution, in O.L.J. Venderborght (ed.), *Proceedings "Acid Deposition and Sulphur Cycle"*, Brussels, pp 109-116.
5. Siepel, H. (1994). *Structure and Function of Soil Microarthropod Communities*, Ph.D. Thesis, Agricultural University, Wageningen, The Netherlands.
6. Joosse, E.N.G. and Verhoef, S.C. (1983) Lead tolerance in Collembola, *Pedobiologia* **25**, 11-18.
7. Kooijman, S.A.L.M. and Metz, J.A.J. (1984) On the dynamics of chemically stressed populations: the deduction of population consequences from effects on individuals, *Ecotox. Environ. Saf.* **8**, 254-274.

8. Schaefer, M. (1989) Soil and litter arthropods of green urban spaces, in *International scientific workshop on soils and soil zoology in urban ecosystems as a basis for management and use of green/open spaces*, Report of the UNESCO Programme on Man and the Biosphere (MAB).
9. Stamou G.P. and Argyropoulou M.D. (1995) A preliminary study on the effect of Cu, Pb, and Zn contamination of soils on community structure and certain life history traits of oribatids from urban areas, *Experim. Appl. Acarol.* (in press)
10. Stamou, G.P. (1981). *Population Dynamics of Achipteria cf. italicus in the Soil Subsystem of an Oakwood in Holomon Mt.* (in Greek), Ph.D. Thesis, University of Thessaloniki, Greece.
11. Argyropoulou M.D. and Stamou G.P. (1992) Respiratory activity of the collembolan *Onychiurus meridiatus, J. Insect Physiol.* **39**, 217-222.
12. Stamou G.P. and Iatrou G.D. (1993) Studies on the respiratory metabolism of *Glomeris balcanica* (Diplopoda, Glomeridae), *J. Insect Physiol.* **39**, 529-535.
13. Stamou G.P., Asikidis, M.D., Argyropoulou, M.D and Iatrou, G.D. (1994) Respiratory responses of oribatid mites to temperature changes, *J. Insect Physiol.* **41**, 229-233.
14. Stamou G.P and Asikidis, M.D. (1992) The effect of certain biotic factors on the demographic parametes of *Scheloribates cf. latipes* (Acari: Oribatida), *Pedobiologia* **36**, 351-358.
15. Levins R. (1968) *Evolution in Changing Environments*. Princeton University Press, Princeton, USA.
16. Terano, T., Asai, K. and Sugeno, M. (1992) *Fuzzy Systems Theory and Applications*, Academic Press, New York, USA.

ESSENTIAL CRITERIA FOR SELECTING BIOINDICATOR SPECIES, PROCESSES, OR SYSTEMS TO ASSESS THE ENVIRONMENTAL IMPACT OF CHEMICALS ON SOIL ECOSYSTEMS

C.A. EDWARDS, S. SUBLER, S.K. CHEN, AND
D.M. BOGOMOLOV
The Ohio State University Department of Entomology and Environmental Science Program
1735 Neil Avenue, Columbus, Ohio, 43210, United States of America

Abstract

The effects of chemicals on particular species or groups of soil organisms, dynamic soil processes, and whole soil systems in microcosms have all been used to evaluate the potential overall environmental impact of chemicals reaching soils, in both laboratory and field experiments.

If **key indicator species or taxonomic groups** are used they should: (i) play a key role in ecosystem processes; (ii) be present in a wide range of soil ecosystems; (iii) exist in large and dominant populations; and (iv) be testable under natural conditions, and (v) have efficient, readily-available and non-laborious methods of assessing populations. Such species/groups could be: nematodes, earthworms, mites, springtails, or carabid beetles.

To select **dynamic soil processes** that can evaluate the potential impact of chemicals on soil systems, such processes must: (i) be critical to the functioning of soil ecosystems; (ii) have well-established methods for assessing changes in their function; and (iii) produce data that are readily interpretable in terms of environmental impact. Such candidate processes include: rates of organic matter breakdown, soil respiration, microbial growth and turnover, soil enzyme activity, and rates of change in nutrient status.

Finally, if a **laboratory microcosm assessment** is used to predict the environmental impact of chemicals on soil ecosystems in the field, it must: (i) duplicate the functions of common soil ecosystems; (ii) have physical, chemical, and biological characteristics similar to a typical field soil, (iii) be of appropriate size to be replicated and handled adequately; (iv) involve a suite of measurements which provide an overall representation of ecosystem function; and (v) provide data easily interpretable in terms of soil pollution of field soils.

N. M. van Straalen and D. A. Krivolutsky (eds.), Bioindicator Systems for Soil Pollution, 67–84.

The merits and demerits of the three types of bioindicators are discussed, an innovative and broadly representative microcosm methodology described, and some typical results for the effects of fungicides and heavy metals in soil microcosms are summarized and discussed.

1. Introduction

The maintenance of soil quality, fertility, and structure is essential for sustainable agricultural production and for the protection and maintenance of the biodiversity and integrity of terrestrial ecosystems. Central to achieving this aim is the need for a vastly improved understanding of the potential effects of chemical contaminants on the structure and function of ecosystems. Chemical contaminants of soils can exert their effects directly, through toxicity to soil organisms, or indirectly, by altering species interactions and by disrupting soil food webs. Ultimately, these effects can interfere with soil processes that may be important in the regulation, flow and internal cycling of carbon and nutrients in ecosystems (Parmelee et al. 1993b; Edwards and Bohlen 1995; Tu and Miles 1976). The effects of chemicals on particular species or groups of soil organisms, dynamic soil processes, and more rarely on whole soil systems, have all been used to evaluate the potential overall environmental impact of chemicals reaching soil, using both laboratory and field experiments.

In practice, soil organisms and carbon and nutrient cycling processes are so intimately linked that the effects of chemicals on one of these aspects are likely to impact on the others. Furthermore, the intensity and duration of the environmental effects of chemicals may often depend strongly upon those processes that influence the activity, persistence, and movement of chemical contaminants though a soil ecosystem and into soil animals and plants. Soil organisms and plant roots, through their interactions with soil chemical and physical processes, and their direct and indirect roles in degradation of chemicals, may influence the bioaccumulation and persistence of such chemicals in soils (Edwards 1992, 1994a; Tu and Miles 1976). Because of the complexity of soil ecosystems, and the tight linkages between soil organisms and soil processes, it is difficult to assess the influence of chemical contaminants on soil ecosystems without considering their potential interactive effects. For this reason, the battery of single-species or single-process tests that have been used to date in soil ecotoxicological research and for environmental risk assessment usually fall short of providing data which can predict effectively the environmental consequences of chemical contamination in the field (Van Voris et al. 1985). Field studies, although perhaps providing realistic conditions for ecotoxicological assessment, are usually limited severely by high soil variability, inadequate controls, poor reproducibility, and exorbitant costs (Edwards 1989a, 1992).

In recent years, terrestrial ecologists have used simplified model ecosystems, or microcosms, as tools to investigate species interactions and ecosystem processes. For example, soil microcosms have been used successfully to elucidate relationships between soil microorganisms and micro- and macro-invertebrates (Anderson and Ineson 1984), to investigate the role of roots on food-web structure and organic matter decomposition (Parmelee et al. 1993a), and to examine the effects of soil and litter-dwelling organisms on organic matter decomposition and nutrient cycling processes (Teuben 1991). However, with few exceptions this work has focused on specific interactions or processes occurring under ideal or `natural' conditions, and has not addressed the overall response of ecosystems to perturbations, such as the addition of potentially toxic chemicals to the soil.

In this paper, we review the use, merits and demerits of key indicator species, taxonomic groups and functional guilds, in the assessment of the impact of chemicals on soil ecosystems. We shall also discuss the use and value of dynamic soil processes as indices of the effects of chemicals on microorganisms and selected groups of soil inhabiting invertebrates, as well as their effects on overall ecosystem function. Finally, we shall review the use of model ecosystems and give examples of the results of an innovative integrated microcosm assay.

2. Key Indicator Species, Taxonomic Groups or Functional Guilds

There is an enormous body of scientific literature documenting the effects of chemicals on soil microorganisms and soil-inhabiting invertebrates (Edwards and Thompson 1973; Thompson and Edwards 1974; Edwards 1989). A number of single species tests that are applicable to assessing the effects of contaminants on soil microorganisms have been described (Edwards 1989). Most of these tests involve culturing individual species of microorganisms, mixing the contaminant with the culture medium, and assessing the effects on growth. Such techniques are commonly used to assess the influence of contaminants on those microorganisms that are active in the nitrogen and carbon cycles (ammonifying, nitrifying, denitrifying, proteolytic and amylolytic bacteria), but they may have limited field relevance. The assessment of the effects of contaminants on mixed species of microorganisms in soil is much more difficult. The microbes have to be extracted from soil in suspension and then cultured; their numbers are so large that only subsamples can be counted; moreover it is difficult to differentiate between living and dead microorganisms, even with tools such as fluorescence and coloration techniques (Edwards 1989). It is virtually impossible to relate the results of such tests to actual exposure of the microorganisms to chemicals in field soils, or to interpret how these effects would influence the functioning of soil processes.

Numerous single-species tests for assessing the effects of contaminants on soil-inhabiting invertebrates have been developed. These have ranged through: injection of test chemicals into invertebrates (Stringer and Wright 1976); topical application of chemicals to invertebrates (Edwards and Thompson 1973); immersion of invertebrates in dilute solutions of chemicals (LeBrun et al. 1981); exposure of invertebrates in contaminant-treated natural soils (Karnak and Hamelink 1982; Lofs-Holmin and Bostram 1988) or in contaminant-treated artificial soil (Edwards 1983, 1984; Edwards and Bohlen 1992). Some workers have also based their conclusions on assessing the effects of contaminants on species within intact populations and communities of invertebrates in the field.

2.1 CRITERIA

It is essential that there be a critical selection of the taxa or key functional groups that are used for soil ecotoxicological testing, hence it is important to list the criteria for the selection of a particular method. The indicator species, taxonomic group or functional guild that is used for ecotoxicological assessment should fulfill the following criteria in that it:

- Must play a key role in the soil ecosystem
- Must be present in a wide range of soil ecosystems
- Must exist in large and dominant populations
- Must be testable under natural or simulated laboratory conditions
- Must have efficient, non-laborious methods for extraction of organisms from soil and of population assessment

2.2 TAXONOMIC GROUPS

The taxonomic groups of microorganisms that have been used most in tests of the effects of chemicals on microorganisms include many species of soil bacteria, particularly *Azotobacter* (Grossbard 1976), *Clostridium* and *Rhizobium* (Anderson 1978), fungi (Bollen 1984), actinomycetes (McCann and Cullimore 1979) and algae (Domsch 1983).

Many species of soil-inhabiting invertebrates have also been used in laboratory toxicity tests. However, the major taxa that have been involved commonly in such tests have tended to be larger taxonomic groups, including nematodes (Parmelee, 1993b), earthworms (Edwards and Bohlen 1992), carabid beetles (Edwards 1979), spiders (Maefait 1995), and woodlice (Laskowski et al. 1995). Microorganisms and invertebrates that are used in laboratory toxicity studies should be able to be reared easily in the laboratory, have a high rate of survival in laboratory culture, have simple food requirements, and have a high reproductive rate.

It seems appropriate to illustrate the use of such taxa groups in soil ecotoxicology by providing examples and data from our own research. We will discuss results based on testing using earthworms and carabid beetles.

2.2.1 *Earthworms*

There have been many studies using earthworms as the key indicator species for assessing the effects of contamination on soil ecosystems. The reason for this selection is because earthworms are very important representatives of the soil fauna which have a key role in facilitating the breakdown of organic matter and release of nutrients in soils. They are also important in maintaining soil structure, porosity, aeration and drainage. Since their major source of nutrition is microorganisms, particularly soil fungi, they can demonstrate indirectly many of the effects of contaminants on microbial communities. Earthworms were chosen as key indicator organisms of environmental pollution by the Organization for Economic Cooperation and Development (OECD) and also the European Economic Community (EEC) (now the European Union) to develop a standardized ecotoxicological test (Edwards 1983, 1984).

A standardized artificial soil was designed to simulate a medium loam soil in terms of main structural and physical characteristics, particularly adsorptive capacity. This was used to eliminate the inherent variability in characteristics of field-collected soil. The earthworm species *Eisenia fetida* was chosen as the test organism, because it could be cultured readily in large numbers. A very precise protocol for the laboratory assay was prepared, based on extensive preliminary experimental testing. The details of the test are summarized in Table 1.

TABLE 1. Summary of the standardized ecotoxicological earthworm test (Edwards 1983).

Formula for Artificial Soil

10% Sphaghum peat (pH 5.5-6.0, finely ground)
20% Kaolinite clay (preferably with more than 50% kaolinite)
69% Industrial quartz sand (dominant fine sand > 50%, particle size 0.05-.2 mm)
1% Calcium carbonate ($CaCO_3$ pulverized and chemically pure) to bring pH of mixture to 6.0

Test Method

One litre containers with lids
Artificial soil based on above formula
Chemical sprayed and thoroughly mixed into soil
Soil moisture: 25-42% dry weight of soil added
Test period: 14 Days
Test temperature: 20° C in continuous light (400-800 LUX)

Test Organisms and Measurements

Test species *Eisenia fetida* > 2 months old with clitellum
Ten earthworms per container
No food provided
Measurements: mortality (missing or no movement)
Measurements: number of cocoons

Three unknown chemicals labeled A, B and C, and a standard chemical, together with a detailed testing protocol, were sent to 20 toxicological laboratories world-wide. Each laboratory was asked to assay the effects of the three chemicals on earthworms and provide an L.C. 50 (with confidence limits) for each chemical. This 'ring test' was repeated in two consecutive years. In the first year, the unknown chemicals sent out were pentachlorophenol (PCP), carbaryl and trichloracetic acid (TCA), and the standard chemical was copper sulphate. In the second year, the three unknowns were potassium bromide, PCP and chlordane and the standard chemical was chloracetamide. The data for the toxicity of chloracetamide from the 20 laboratories and L.C. 50 values for the seven chemicals are summarized in Figure 1.

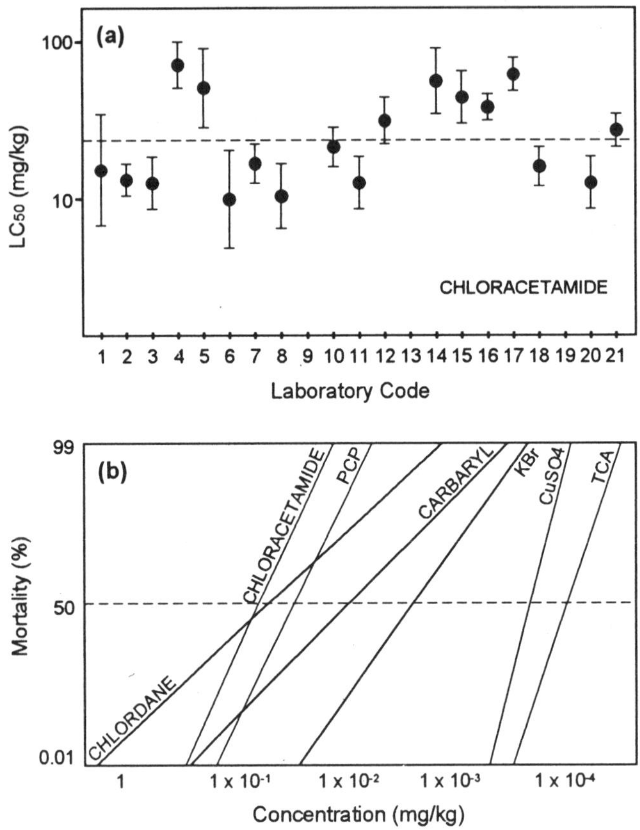

Figure 1. (a) Toxicity of chloracetamide to earthworms determined using the standardized earthworm test at 20 different laboratories, and (b) comparative toxicity of seven chemicals to earthworms from the 'ring test.'

2.3.2 *Carabid beetles*

Many workers have used carabid or staphylinid beetles as key test organisms for assessing the effects of soil contamination because of the importance of these insects as predators of a wide range of crop pests. For instance, the susceptibility of carabids to a wide range of pesticides has been tested in both laboratory and field experiments. We tested the toxicity of a range of organophosphate insecticides to carabid beetles in both the laboratory and field. The insecticides tested were chlorfenvinphos, diazinon phorate, and parathion.

For the laboratory assays, 5 µl of acetone solutions of the insecticides, at appropriate concentrations, were applied with a micropipette to the ventral surface of the abdomen of each beetle. After treatment at each dose, groups of five beetles were kept in separate 9.5 cm diameter crystallizing dishes without food or water. A control group of twenty-five beetles was treated with acetone only. Dead beetles were counted after 48 h, both dead and moribund beetles being termed `dead'. None of the beetles treated only with acetone died.

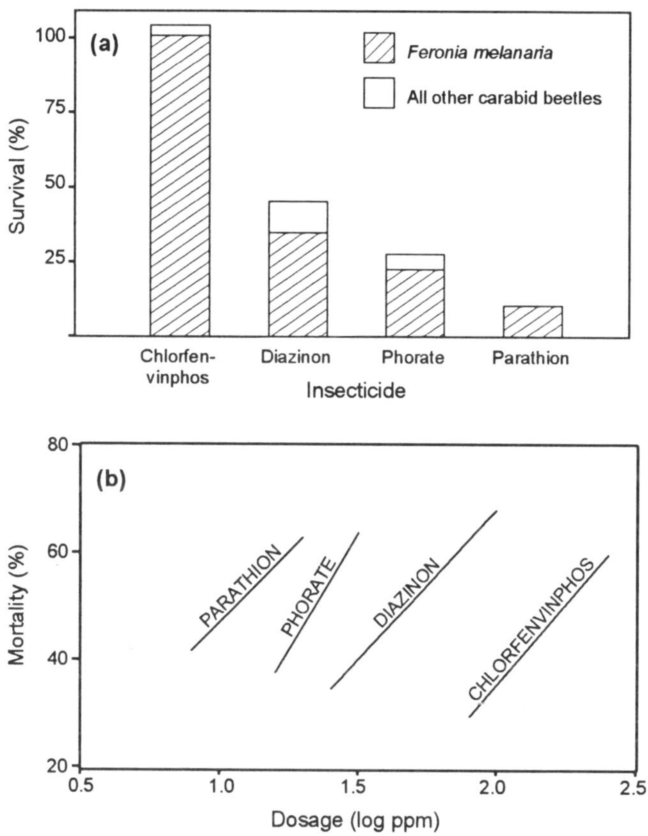

Figure 2. Toxicity of organophosphate insecticides to carabid beetles; (a) field, (b) laboratory.

For field testing, polyethylene barriers buried 15 cm deep in soil and standing 40 cm above the soil surface were erected around the perimeter of plots 8.0 x 5.5 m., separated by paths 4.5 m wide. Plots were arranged with four replicates of each treatment in randomized blocks. The barriers were erected when the beetles first became active in late May or early June, and eight 10 cm diameter pitfall traps, each marked by a cane 1 m high, were placed in each plot. The catches in the pitfall traps were examined and identified to species every 2 days, and the beetles released again on the same plots on which they were caught. After 1 week of trapping, insecticides were applied as either surface sprays or broadcast granules. Thereafter, catches were recorded every 2 days until beetle activity ceased. Results are summarized in Figure 2 in terms of the total numbers of carabids caught relative to untreated plots (a) and, for each insecticide treatment, as the percentage mortality as a function of dosage (b).

2.4 FUNCTIONAL GUILDS

A number of workers have used functional guilds of selected organisms to assess toxicity of chemicals to soil ecosystems. These have included plant parasitic, fungivorous and bacterivous, omnivorous and predatory nematodes (Parmelee et al. 1995) and staphylinid beetles with different feeding habits (Bohac 1995).

3. Dynamic Soil Processes

Organic matter decomposition and nutrient cycling processes in soils involve biochemical, chemical, and physicochemical reactions that are mediated by microorganisms, soil invertebrates, and plant roots. All such biochemical reactions are catalyzed by enzymes specific to particular biochemical pathways. The primary pathways for input of energy and many nutrients to soil ecosystems are through plant litter and roots. Rates of decomposition of organic litter are regulated by soil microbial and faunal interactions, and these in turn determine the trophic structure of soil communities (e.g., Hendrix et al. 1986; Beare et al. 1992; Swift et al. 1979). The cycling of nutrients, particularly nitrogen, is associated intimately with organic matter decomposition, since nutrients are often concentrated within organic materials, and because organic matter provides the energy (organic carbon) necessary to drive the biochemical transformations between organic and inorganic nutrient pools. Disruption of trophic interactions among soil organisms can result in large and long-lasting effects on nutrient cycling processes, and therefore on soil quality and above-ground productivity (Hendrix et al. 1986).

3.1 CRITERIA FOR SUITABLE SOIL PROCESSES FOR ECOTOXICOLOGICAL TESTING

The soil processes that have the greatest potential in assessment of the effects of contaminants on soil ecosystems include the following:

- Organic matter breakdown
- Soil respiration
- Changes in biochemical function (soil enzyme activity)
- Nutrient cycling

Whichever of these soil processes is chosen as an index of soil activity in ecotoxicological testing, it must conform to a number of essential criteria; it must:

- Be critical to the function of soil ecosystems
- Have well-established methods for assessing changes in rates or function
- Produce data readily interpretable in terms of overall environmental impact

We shall discuss the value of these processes in ecotoxicological testing and discuss results from organic matter breakdown and respiration studies from our own research.

3.2 EXAMPLES OF PROCESSES

Any of the four processes described above can be used. We can summarize them as follows:

3.2.1 *Organic Matter Breakdown*

The influence of chemical contaminants on rates of organic matter decomposition in soil has been assessed by exposing various organic materials to attack by soil organisms in uncontaminated soil as a control, and comparing the rate of mass loss of similar materials in soil treated with different doses of contaminants (Edwards 1979). The rates of decomposition of organic plant litter following the application of herbicides in agroecosystems has been of particular concern (Harden et al. 1993; Hendrix and Parmelee 1985).

3.2.2 *Soil Respiration*

Soil respiration can be measured readily in the soil using several techniques, most depend upon the absorption of carbon dioxide by an alkali. Substrate-induced respiration (SIR), based on measurements of CO_2 production following amendment with a glucose solution can be also be used. These techniques can be used with equal facility in both laboratory and field assays (Jenkinson and Powlson, 1976).

3.2.3 *Changes in Biochemical Function (soil enzyme activity)*

The determination of the activity of free enzymes in soil and soil organisms (Bitton and Koopman 1992; Frankenberger and Dick 1983) has good potential for ecotoxicological testing in soils. Some workers have investigated the effects of contaminants on enzymatic activities in soils, especially their effects on enzyme inhibition or synthesis; with particular attention to hydrogenases, phosphatases, ureases and glucosidases (Hickisch et al. 1984; Cervelli et al. 1978).

3.2.4 Nutrient Cycling

A number of workers have assessed the effects of contaminants on nitrogen transformations in soils. This has ranged from plate counts of *Rhizobium* cells to effects on the rate of nodulation of legumes (Anderson 1978). The effects of contaminants on ammonification (the release of ammonia from organic matter) has been studied after addition of a nutrient source such as lucerne meal to soil. The effects of insecticides on nitrogen transformations have been reviewed (Tu and Miles 1976). As with the microbial assays, the resulting information may be difficult to interpret with respect to interactions of soil organisms and soil processes in terms of predicting process-level effects in a field setting.

4. Laboratory Microcosms

4.1 THE USE OF MICROCOSMS IN ECOTOXICOLOGICAL STUDIES

In recent years, renewed interest has developed in the use of microcosm techniques for ecotoxicological assays (Sheppard 1994; Van Voris et al. 1985). Such microcosms have consisted of soil units containing multiple biotic species, and have ranged in size from a few grams of soil to as large as a meter in diameter. They can consist of mixed cultures of organisms and artificial substrates (Ausmus et al. 1979), mixed soil containing organisms (Lichtenstein 1980; Edwards and Bohlen 1992) or intact cores from the soil ecosystem that they represent (Van Voris 1985; ASTM 1988).

The attraction of microcosm tests is that they can assess the impact of a contaminant at different levels of biological organization and produce data that is much more relevant to field situations. However, some currently-used microcosm tests, although containing multiple organisms, assess the effects of the contaminant on only one species. For instance, the method described by Lichtenstein (1980) and the one described in the American Society for Testing and Materials (ASTM) Annual Book of Standards (1988) involves only one planned test species; i.e. the plant cover in intact soil cores. Although many other species of organisms are probably present in the microcosm, only the plant response, and nutrient and trace element leaching losses, were recorded in the test. Although some authors have argued strongly for the value of microcosms and multispecies testing for assessing the impact of contaminants on whole soil ecosystems (Cairns, 1985, 1986), many regulatory authorities still depend upon and promulgate single-species testing.

Sheppard (1994) discussed the cost and benefits of single species tests compared with microcosm tests and concluded that the former are cheaper, but the value of the information derived from microcosms can greatly outweigh the additional costs of the multiple tests. If a microcosm test is fully standardized and reproducible, it can produce a large amount of information at relatively low cost and with little need for additional single-species tests. It is clear that no single test can assess the effects of a contaminant in all settings, whether based on species or process testing or microcosms. However, microcosm testing does allow a better overall interpretation of potential impacts of a contaminant upon the trophic structure of soil communities, and useful information on the effects of contaminants on ecological functions. There seems to be little doubt that extrapolation of the results of a microcosm test to the field is much easier than with the results of single species tests (Cairns and Pratt 1987). For instance, Verhoeff (1992) measured enzyme activities, respiration and nutrient release in decomposing plant litter in 0.16 L and 2.1 L microcosms and found very good correlation with field data.

The question of scale of microcosms was addressed by Wright and Coleman (1993) who monitored the effects of chemicals on nematode populations in replicated cores of 15, 49, 180, and 7600 cm^3. They concluded that for such small organisms there were serious ecological drawbacks in the larger-sized microcosms, because nematodes in the higher trophic levels were often lost.

Microcosm tests that have been developed to date include one, discussed previously, that was a small-scale system that assessed the effects of contaminants on communities of soil micro- and meso-arthropods and nematodes (Parmelee et al. 1993b). Other workers have used samples as small as 4 g soil to assess the effects of contaminants. The system described by Van Voris (1985) which was modified as ASTM method Vol. 11.04, Section 11 (1988), was much larger and involved intact cores 17.5 cm diam. x 60 cm deep. Kuehle (1988) described an even larger system (NATEC) which consisted of undisturbed soil cores 7.2 cm diam. x 43 cm deep planted with four different crops. Kappers and Manger (1990) and Kappers and van Estoroek (1987) also used similar-sized cores to assess the effects of chemicals on nematodes.

Nearly all of the microcosms described in the literature to date are large, and use intact soil cores. Although these can simulate field conditions well, and may be especially useful for leaching studies, they create major problems in reproducibility because of: soil variability, particularly for large microcosms, the lack of contact between soil organisms and the contaminant, and the relatively short time-span of many tests. Moreover, they typically measure a limited range of responses to the contaminant.

4.2 AN EXAMPLE OF AN INTEGRATED MICROCOSM APPROACH

We have designed a small integrated microcosm assay, which can provide organismal and system-level information on the effects of toxicants (Edwards et al. 1994b). This microcosm is illustrated in Figure 3. Each microcosm consists of a core of sieved, field-collected soil, packed gently in a plastic cylinder, with wheat seedlings planted into the soil. The outer cylinder of the microcosms is 5 cm (inside diam.) x 15 cm high made of commercially available high density polyethylene (HDPE) pipe. A small quantity of organic material (chopped wheat straw) within a small cylinder of fiberglass screen material is inserted into the surface of the microcosm soil, so that measurements of the rates of decomposition of the organic material can be made.

In addition to the microorganisms and invertebrates, (including nematodes) that are present in the fresh field-collected soil, two small earthworms of the species *Aporrectodea tuberculata*, or other species, are added to each microcosm. The earthworms are macrofaunal integrators of the effects of chemicals on soil microbial and food web processes, and serve as biomarkers of the bioaccumulation and toxicity of contaminants. Upon termination of the microcosm experiments, earthworm biomass, tissue carbon and nitrogen concentrations, and tissue contaminant concentrations are measured.

Wheat seedlings are planted in the surface of the microcosm soil. The importance of including plants in soil microcosms is two-fold; (i) plant roots serve as a source of organic carbon in the soil, playing a fundamental role in below-ground food webs, and (ii), plant roots withdraw nutrients, as well as chemical contaminants, from the soil, altering soil solution concentrations and affecting processes controlled by positive or negative feedbacks. Germination success, growth, and nutrient and contaminant concentrations of the wheat seedlings are measured.

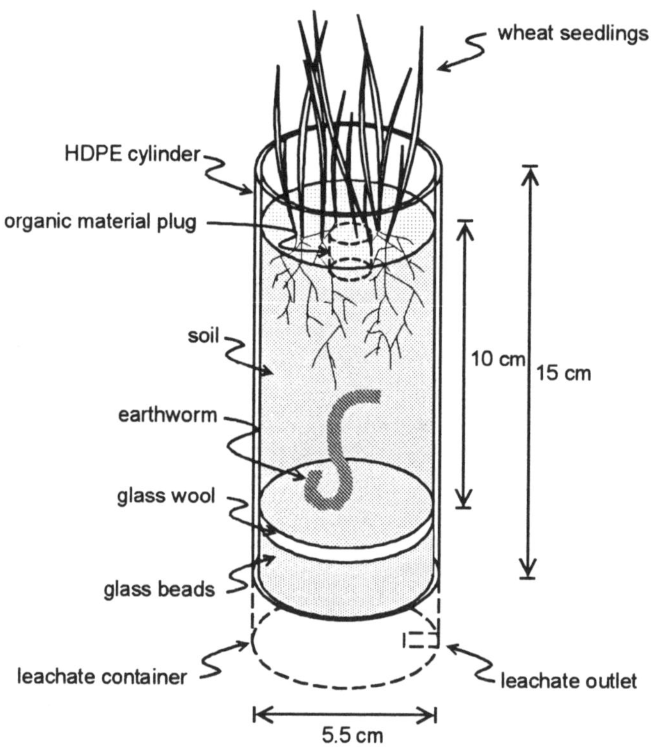

Figure 3. Diagram of an integrated soil microcosm for ecotoxicological testing

In the bottom of the microcosm cylinder is a layer of glass beads separated from the soil core by a thin layer of glass wool. The glass beads are held in place by polyethylene screen secured to the bottom of the cylinder. The glass wool and beads serve to retain soil in the microcosm while allowing free passage of soil leachate, and act as a partial barrier to prevent root growth out of the core bottom. The entire microcosm sits on top of a glass container for the collection of leachate so that leaching losses of nutrients may be determined.

Because of the small size and relative simplicity of the soil microcosms, large numbers of replicates can be handled in individual experiments. A subset of microcosms can be sampled destructively after different periods throughout each experiment (for example, 0,2,7, 14, 28, and 56 days after chemical treatment). At the end of each period, soil from each microcosm is removed and mixed gently, then subsampled to determine the biomass and numbers of soil microbes, microarthropods, nematodes, and earthworms, and to determine soil enzyme activity (dehydrogenase, cellulase, urease and phosphatase), rates of organic matter decomposition, and nutrient status. Plant shoot and roots are separated and analyzed for biomass and tissue nutrient concentrations.

We used microcosms similar to those described above to investigate the influence of two chemicals, copper (a heavy metal) and captan (a fungicide), on soil organisms and processes. These chemicals were chosen because of their importance as soil contaminants due to industrial and urban pollution and sewage sludge application, (copper), or their widespread use as agricultural pesticides (captan). Additionally, these chemicals differ in their direct toxic effects to soil organisms: copper is broadly toxic to many soil organisms and plants, at higher concentrations; captan, although having some non-target effects, is acutely toxic to organisms at the base of the decomposer food web (fungi).

The effects of captan on substrate-induced respiration and net nitrogen mineralization in the soil microcosms are presented in Figure 4. Substrate-induced respiration, which provides an index of microbial activity, was suppressed by treatment with captan at typical field rates after 28 and 56 days, relative to untreated controls. Clearly, short-term soil incubations are likely to miss such significant long-term responses and underestimate the impact of the captan on overall microbial activity. The dynamics of net nitrogen mineralization in the soil over the 56 day test period were also influenced strongly by captan treatment, alternating between increased net immobilization and mineralization relative to untreated microcosms. In field settings, this may have implications for the timing of nitrogen availability for plant uptake, or for loss of nitrogen through leaching, volatilization, or denitrification.

The effects of varying concentrations of copper on nematode populations and organic litter decomposition in soil microcosms are presented in Figure 5. In this case, the process-level measure was more sensitive to copper contamination, showing significant differences from untreated controls at 200 ppm, whereas nematodes responded significantly only at the highest concentration (800 ppm).

4. Conclusions

We have reviewed the criteria that influence the use of key indicator species, taxonomic groups, functional guilds, dynamic soil processes and soil microcosms, in the

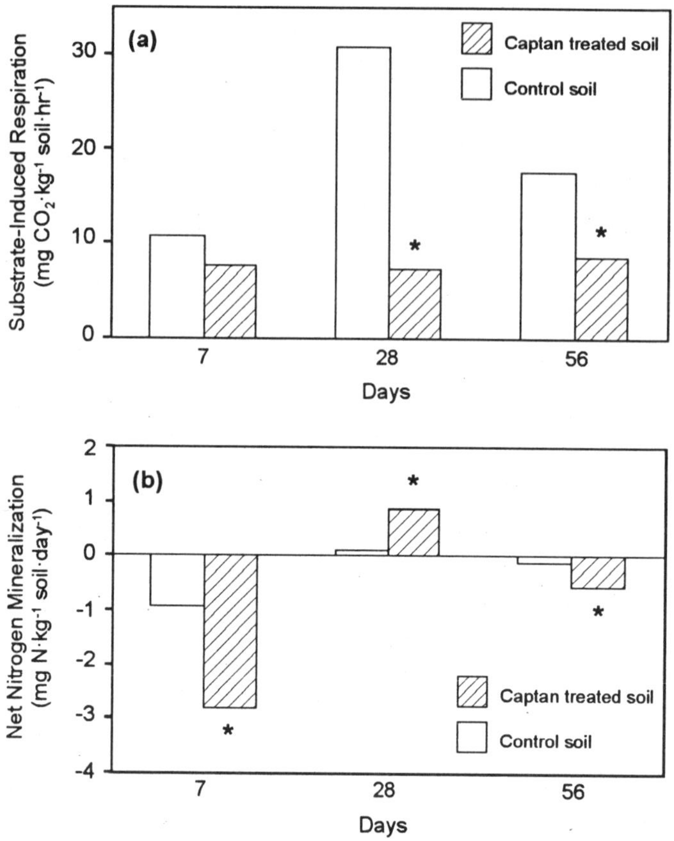

Figure 4. The effects of the fungicide captan on substrate-induced respiration (a) and net nitrogen mineralization (b) in soil microcosms (Source: Chen et al. unpublished). Asterisks indicate significant difference from control.

assessment of the impacts of chemicals on soil ecosystems, referring to both laboratory and field testing techniques. Our overall conclusions are that: any one single-species or process test is inadequate to make an accurate forecast of potential environmental hazards; we suggest that if a species-based assessment is used, at least one micro-invertebrate or meso-invertebrate, together with a macro-invertebrate assay is needed. If only one soil process is used for forecasting, then organic matter breakdown is probably the best single index, but data on enzyme activities and nutrient cycling are also valuable, either alone or as supporting information.

We conclude that an integrated soil microcosm of the type we describe can probably provide the best information for overall predictions of potentially serious impacts of chemicals on populations of key soil organisms, on the interactions between

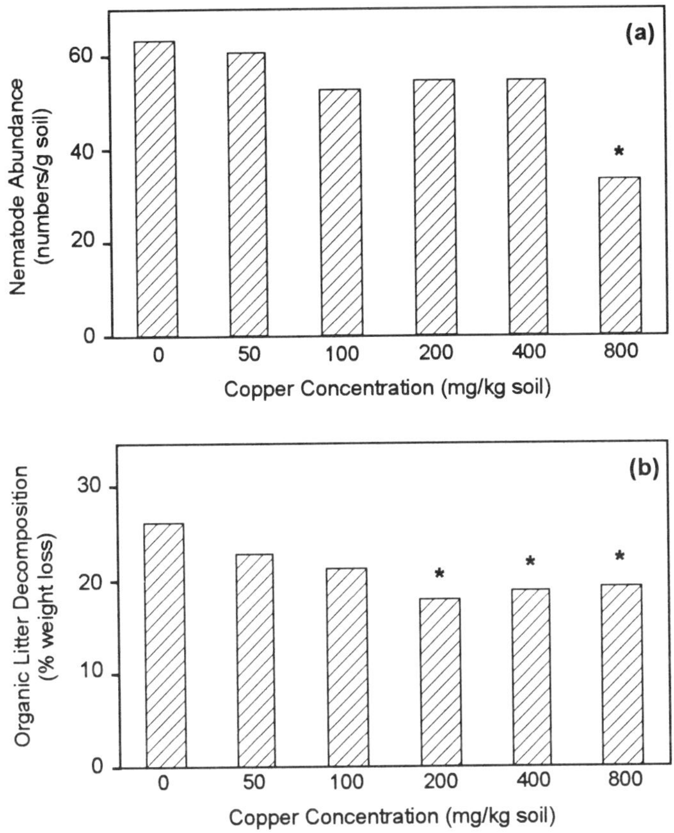

Figure 5. Nematode abundance (a) and percentage weight loss of organic litter (b) in microcosm soils 40 days after treatment with varying concentrations of copper (Source: Bogomolov et al. unpublished). Asterisks indicate significant difference from untreated control.

soil organisms, and on important soil processes which function to maintain soil structure and fertility, as well as the biodiversity and integrity of terrestrial ecosystems.

5. References

1 Anderson, J. M. and P. Ineson. 1984. Interactions between micro-organisms and soil invertebrates in nutrient flux pathways of forest ecosystems. In J. M. Anderson, A. D. M. Rayner and D. W. H. Walton (eds), Invertebrate-Microbial Interactions, Cambridge University Press, pp.59-88.

2. Anderson, J. R. 1978. Pesticide effects on non-target soil microorganisms. In Pesticide Microbiology (Ed.) Hill, I R.,a nd Wright, S. J. L. Academic Press, London, pp. 313.

3. ASTM. 1988. Standard guide for conducting a terrestrial soil-core microcosm test. American Soc. for Testing and Materials, ASTM Standard E 1197-87.

4. Ausmus, B. S., Kimbrough, S., Jackson, D. R., and Lindberg, S. 1979. The behavior of hexachlorobenzene organic soil pollutant in pine *Pinus* forest microcosms: transport and effects on soil processes. Environ. Pollut. 20(2):103-11.

5. Beare, M. H., Parmelee, R. W., Hendrix, P. F., Cheng, W., Coleman, D. C., and Crossely, D. A. Jr. 1992. Microbial and faunal interactions and effects on litter nitrogen and decomposition in agroecosystems. Ecol. Monogr. 62:569-591.

6. Bohac, J. 1995. Staphylinid beetles in bioindication and biotesting. Proc. NATO Adv. Res. Workshop (this volume).

7. Bollen, G. J. 1979. Side-effects of pesticides on microbial interactions. In B. Schippers and W. Gams (eds), Soil-Borne Plant Pathogesn, Academic Press, London, pp. 451.

8. Buttan, G. and B. Koopman. 1992. Bacterial and enzymatic bioassays for toxicity testing in the environment. Rev. Env. Contam. Toxicol. 125: 1-22.

9. Cairns, J. Jr, and Pratt, J. R. 1987. Ecotoxicological effect indices: a rapidly evolving system. Water Sci. Technol. 19:1-12.

10. Cairns, J. Jr. 1985. Multispecies toxicity testing. Pergamon Press, Oxford. 259 pp.

11. Cairns, J. Jr. 1986. Community toxicity testing. American Soc. for Testing and Materials, ASTM Spec. Tech. Publ. 920, Philadelphia, PA, 345 pp.

12. Cervelli, S, P., Nannipieri, and P. Sequi. 1978. Interactions between agrochemicals and soil enzymes. In R. G. Burns (ed), Soil Enzymes, Academic Press, New York, pp. 251-293.

13. Domsch, K. H. 1983. An ecological concept for the assessment of side-effects of agrochemicals on soil microorganisms. Residue Rev. 86: 65.

14. Edwards, C. A. 1979. Tests to assess the effects of pesticides on beneficial soil organisms. Proc. Umwelt. Sem. Freie Universitet, Berlin, pp. 240-253.

15. Edwards, C. A. 1983. Development of a standardized laboratory method for assessing the toxicity of chemical substances to earthworms. Report EUR 8714 En. Environment and Quality of Life. Commission of the European Communities 141 pp.

16. Edwards, C. A. 1984. Report of the second stage in development of a standardized laboratory method for assessing the toxicity of chemica substances to earthworms. Repot EUR 9360 EN. Environment and Quality of Life. Commission of the European Communities 99 pp.

17. Edwards, C. A. 1989a. The assessment of the ecological effects of soil pollution by chemicals. Bioiindicatores Deteriorisationis Regionis, Czechoslovakia. Proceedings Vth International Conference, 93-104.

18. Edwards, C. A. 1989b. The impact of herbicides on soil ecosystems. Critical Reviews in Plant Sciences. C.R.C. Press. Cleveland, Ohio 8(3):221-257.

19. Edwards, C. A. 1992. The impact of pesticides on the environment. In D. Pimentel (ed), The Pesticide Question: Environment, Economics and Ethics. Chapter 2, Chapman & Hall, London & New York, 13-46.

20. Edwards, C. A. 1994a. Pesticides as environmental pollutants. In World Directory of Pesticide Control Organizations. 2nd Edition, (Ed.) E. Ekstrom. Crop Protection Publications. Royal Society of Chemistry 1-24.

21. Edwards, C. A., and Bohlen, P. J. 1992. The assessment of the effects of toxic chemials upon earthworms. Reviews of Environemntal Contamination and Toxicology. 125:23-99.

22. Edwards, C. A., and Thompson, A. R. 1973. Pesticides and the soil fauna. Residue Rev. 45:1.

23. Edwards, C. A., S. Subler, and S. K. Chen. 1994b. A microcosm system for evaluating the effects of pesticides on dynamic soil processes and soil organisms. Brighton Crop Protection Conference: Pests and Diseases, 9C-4: 1301-1306.

24. Frankenberger, W. Y. and W. A. Dick. Relationships between enzyme activities and microbial and growth and activity indices in soil. Soil Sci. Soc. Am. J. 47: 945-951.

25. Grossband, E. 1976. Effects of herbicides on the soil microflora. In L. J. Andus (ed), Herbicides, Physiology, Biochemistry and Ecology, Academic Press, London, p. 99.

26. Harden, T., R. G. Joergensen, B. Meyer, and V. Wolters. 1993. Mineralization of straw and the formation of soil microbial biomass in a soil treated with simazine and dinotreb. Soil Biol. Biochem. 25: 1273-1276.

27. Hendrix, P. F. and R. W. Parmelee. 1985. Decomposition, nutrient loss and microarthropod densities in herbicide-treated grass litter in a Georgia piedmont agroecosystem. Soil Biol. Biochem. 17: 421-428.

28. Hendrix, P. F., R. W. Parmelee, D. A. Crossley, Jr., D. C. Coleman, E. P. Odum, and P. M. Groffman. 1986. Detritus food webs in conventional and no-tillage agroecosystems. BioScience 36: 374-380.

29. Hickisch, B., Machulla, G., and Muller, G. 1984. Side effects of herbicides on soil organisms after one and several applications thereof. 3. laboratory experiments: test program, results, obtained for bacterial and fungal spreading density, dehydrogenase activity, and soil respiration after several applications.

30. Jenkinson, D. B., and Powlson, D. S. 1976. The effects of biocidal treatments on metabolism in soil. V. A method for measuring soil biomass. Soil Biol. Biochem. 8(3):209.

31. Kappers, F. I., and Manger, R. 1990. Population dynamics of free-living nematodes in oil contaminated soil during the clean-up with a microbiological restoration technique. Neatologica 36:363.

32. Kappers, F. I., and van Esbroek, M. L. P. 1987. Use of soil organisms in experiment on vulnerability of soil to pollutants. Verslagen en Mededelingen, Commissie voor Hydrologisch Onderzoek TNO 38:161-164.

33. Karnak, R. E., and Hamelink, J. L. 1982. A standardized method for determining the acute toxicity of chemicals to earthworms. Ecotoxicol. Environ. Safety. 6:216.

34. Laskowski, R., E. Pyza, and J. Wojtusiak. 1995. Sublethal toxicity tests for long-living iteroparous invertebrates. Proc NATO Adv. Res. Workshop (this volume).

35. Lebrun, P., De Medts, A., and Wauthy, G. 1981. Comparative ecotoxicology and bioactivity of three carbamate insecticdes on an experimental population of the earthworm *Lumbricus terrestris*. Pedobiologia 21:225.

36. Lichtenstein, E. P. 1980. Fate and behavior of pesticides in a compartmentalized microcosm. In Microcosms in Ecological Research (Ed.) Giesy, J. P. Techn. Info. Cneter, U.S. Dept. Energy, DOE Symposium Series 52 CONF-781101, pp 954-970.

37. Lofs-Holmin, A., and Bostrom, R. 1988. The use of earthworms and other soil animals in peesticide testing,. In Earthworms in Waste and Environmental Management (Ed.) Edwards, C. A. and Neuhauser, E. F., S.P.B. Academic Publ. Co., The Hague, pp. 303-314.

38. Maefait, J. P. 1995. Soil spiders and bioindication. Proc NATA Adv. Res. Workshop (this volume).

39. McCann, A. E. and D. R. Cullimore. 1979. Influence of pesticides on the soil algal microflora. Residue Rev. 1.

40. Parmelee, R. W., J. G. Ehrenfeld, and R. L. Tate III. 1993a. Effects of pine roots on microorganisms, fauna, and nitrogen availability in two soil horizons of a coniferous forest spodosol. Biol. Fertil. Soil 15: 113-119.

41. Parmelee, R. W., J. G. Ehrenfeld, X. Han, and W. Zhu. 1995. Effects of dead pine root inputs on soil fauna and the decomposition of dead roots in a spodosol soil from the New Jersey Pinelands. Acta Zool. Fennica 196: 27-32.

42. Parmelee, R. W., R. S. Wentsel, C. T. Phillips, M. Simini, and R. T. Checkai. 1993b. Soil microcosm for testing the effects of chemical pollutants on soil fauna communities and trophic structure. Environ. Toxicol. and Chem. 12: 1477-1486.

43. Sheppard, S. C. 1994. Toxicity testing using microcosms. In Soil Ecotoxicology (Ed.) G. Bitton, J. Tarradellas and D. Rossel, Lewis Publishers.

44. Stringer, A., and Wright, M. A. 1976. The toxicity of benomyl and some related 2-substituted benzimidazoles to the earthworm *Lumbricus terrestris*. Pestic. Sci. 7:459.

45. Swift, M. J., O. W. Heal, and J. M. Anderson. 1979. Decomposition in terrestrial ecosystems. University of California Press, Berkeley, CA.

46. Teuben, A. 1991. Nutrient availability and interactions between soil arthropods and microorganisms during decomposition of coniferous litter: a mesocosm study. Biol. Fertil. Soil 10: 256-266.

47. Thompson, A. R. and C. A. Edwards. 1974. Effects of pesticides on non-target invertebrates in freshwater and soil. Soil Sci. Soc. Am. Special Publ. 8: 341-386.

48. Tu, C. M. and J. R. W. Miles. 1976. Interactions between insecticides and soil microbes. Residue Rev. 64: 17-65.

49. van Voris, P., D. A. Tolle, M. F. Arthur, and J. Chesson. 1985. Terrestrial microcosms: applications, validation and cost-benefit analysis. In J. Cairns, Jr. (ed), Multipsecies toxicity testing. Pergamon Press, New York, pp. 117-142.

50. Wright, D. H., and Coleman, D. C. 1993. Patterns of surival and extinction of nematodes in isolated soil. Oikos (Copenhagen) 67:563-572.

THE MATURITY INDEX AS AN INSTRUMENT FOR RISK ASSESSMENT OF SOIL POLLUTION

G.W. KORTHALS[1], R.G.M. de GOEDE[2], J.E. KAMMENGA[1] and T. BONGERS[1]

[1] *Department of Nematology, Wageningen Agricultural University, Binnenhaven 10, NL-6709 PD Wageningen, The Netherlands*
[2] *Department of Terrestrial Ecology & Nature Conservation, Biological Station of Wageningen Agricultural University, Kampsweg 27, NL-9418 PD Wijster, The Netherlands*

Abstract

It is widely acknowledged that the impact of contaminants on ecosystems is difficult to predict from single-species laboratory tests. Moreover there is growing evidence that the ultimate effect of a toxicant can be mediated through indirect effects. Studies carried out at the community level provide a meaningful step between single-species laboratory test and ecosystem studies. In order to interpret complex changes in community structure, there is an urgent need to develop new approaches to monitor soil pollution.

The present paper reports on methods to detect various kinds of disturbances by studying changes in the composition of natural occurring nematode communities. We examined wether the Maturity Index (MI), originally developed to monitor colonization and subsequent development of the nematode fauna, can also provide an instrument to measure the effects of soil pollution. It was found that within the Maturity Index concept, most taxa originally scaled in c-p 1 indicate food-rich conditions. Withholding these taxa from the calculation of the MI results in an index (MI2-5) to indicate disturbances resulting from chronic soil pollution, such as heavy metals.

N. M. van Straalen and D. A. Krivolutsky (eds.), Bioindicator Systems for Soil Pollution, 85–93.

1. Introduction

The ultimate goal of ecotoxicology is to detect, predict and monitor the effects of pollutants on ecosystems [1]. To overcome the existing gap between our knowledge of the effects that pollutants have on single species or on ecosystems, there is an urgent need to develop new bioindicator systems with more environmental realism [2]. Studying pollution-induced changes in the structure of natural occurring faunal communities can be an important step to bridge this gap. The advantage of such studies is that they also include mechanisms in which the pollutant indirectly affects the community, like for example via changes in the food availability. This can lead to a better understanding of the role ecological interactions play in toxicological responses of more complex systems [3, 4].

The objective of the present paper is to demonstrate new approaches in the development of bioindicator systems based on changes in the structure of nematode communities. Therefore, first a general overview on the successional stages in the development of faunal communities in general and nematode communities in particular will be given. Next we will test the hypothesis that effects of disturbances can be described as a retrogression in the succession of the nematode community. To test this hypothesis the effects of several kinds of disturbances on the nematode community are studied in terms of changes in the proportion of colonizers and persisters i.e. r- and K- strategists, respectively. Finally, the prospects of the nematode Maturity Index [5] an index based on the relative abundance of colonizers and persisters, as an instrument to discriminate between several types of disturbances of the soil ecosystem, will be discussed.

2. Changes In The Structure Of Communities During Succession.

Faunal communities within an ecosystem are in a continuous process of change. It has been suggested that certain trends can be distinguished during successional changes of ecosystems [6]. Newly formed habitats are invaded by colonizers (I), which results in an initial dominance of rapidly reproducing species (II). Subsequently, more organisms invade the system and the species diversity and structural and functional complexity increases (III). Finally, the system matures and becomes less controlled by external disturbances but more by biotic interactions (IV). In this stage there is a strong selection for species with a high competitive ability (V).

Although it can be debated whether succession is an unidirectional process and comprise predictable changes, there is a long tradition among ecologists to unravel the biotic and abiotic properties which determine the ecological range of the species within a successional sequence. The theory of r- and K-strategies showed that the degree of crowding can be one of the explaining factors. Although the meaning of the r/K concept has broadened, originally it was suggested that r-selected species are characterized by a high population growth in uncrowded populations, whereas K-selected species exhibit a high competitive ability in crowded populations [7]. The r-strategists are the successful colonizers. They have a high rate of population increase and a high dispersal ability. In uncrowded environments, like early stages of succession, these species are in favour. K-strategists show opposite characteristics and are found in more stable habitats, in which they maximize their carrying capacity [8,9].

Theoretically all species can be placed on a r-K continuum of strategies, of which some important characteristics are listed below:

Strategy:	r	K
<u>Community attributes</u>		
Diversity	Low	High
Interspecific competition	Occasional	Frequent
Degree of specialization	Low	High
<u>Species attributes</u>		
Colonization ability	High	Low
Distribution	Wide	Restricted
Parthenogenese	Variable	Low
Length of life	Short	Long
Maturity	Early	Late
Rate of development	Rapid	Slow
Fecundity	High	Low
Population density	Variable	Constant
Capacity for dormancy	Variable	Low

The r-K theory has been debated thoroughly. Most of the criticism was based on the fact that other factors than crowding can explain the diversity in life-history strategies as well [10-12], or that certain species exhibit a dynamic strategy involving shifts between relative r/K-positions along the r/K-continuum [13], or the fact that some pioneer species do have certain characteristics which are not in line with the r/K concept [14]. Despite this criticism there is much empirical support for the r/K concept, which warrants an analyses of the application of this ecological concept in studying changes in the structure of faunal communities within ecotoxicological studies.

3. Changes In The Structure Of Nematode Communities During Succession

A helpful tool in studying changes in the structure of nematode communities during natural succession is the nematode Maturity Index [5]. Based on their ability to colonize new habitats, nematode families were classified on a tentative colonizer-persister (c-p) scale ranging from 1 to 5. Nematode families comprising species that rapidly increase in number in early stages of succession, were considered as colonizers and received a low c-p value. They have similar characteristics as r-strategists. In general colonizers live in unstable habitats. Species of the families Rhabditidae, Panagrolaimidae and Diplogasteridae represent typical colonizers. The persisters among the nematodes are comparable with K-strategists and they generally live in habitats with a long durational stability. The most extreme persisters are found among the families Nygolaimidae, Thornematidae, Belondiridae, Actinolaimidae and Discolaimidae. During the last couple of years, further information on this MI-concept has become available [15-17].

The MI is calculated as the weighted mean of the c-p values assigned to the constituent nematode families (and genera and species they contain), such that:

$$MI = \frac{\sum_{i=1}^{n} (v(i) \, . \, a(i))}{\sum_{i=1}^{n} a(i)}$$

where v(i) is the c-p value assigned to taxon i and a(i) is the abundance of taxon i in a sample. If, for example, a community exists of 10 Rhabditidae (c-p value=1), 10 Diplogasteridae (c-p value=1), 30 Cephalobidae (c-p value=2) and 50 Dorylaimidae (c-p value=5) the MI is {(10+10)*1 + (30)*2 + (50)*5}/(10+10+30+50) = (330/100) = 3.3.

Following this classification of nematode taxa into c-p groups it is expected that early stages in the succession of a habitat will be characterized by relatively low MI values. The first colonizers comprise species with the characteristics of r-strategists and thus with low c-p values. Subsequently, during further development of the habitat the MI will increase, because of increasing numbers of K-strategists belonging to the higher c-p values. Such pattern of nematode fauna development was observed in several studies comprising natural successional seres [18, 19].

4. Changes In The Structure Of Nematode Communities And MI After Disturbances

Natural succession of habitats coincides with patterns of nematode community development, generally resulting in an increasing value of the Maturity Index. However, the process of succession can be influenced by external disturbances, resulting in a retrogression or an arrestment of the succession of habitats [20-22]. If environmental disturbances also lead to a retrogression or arrestment of the development of nematode communities, then this should result in decreased MI values.

In terms of the c-p group classification within the MI-concept, two kinds of changes in the composition of the nematode fauna after a disturbance were suggested [5, 23]. A type I response is found when disturbances result in increased numbers of taxa with low c-p values (1-2), whereas taxa from the higher c-p groups hardly respond [see e.g. 24]. A second type of response (response type II), is found when disturbances result in an absolute decrease among most taxa, but in particular those with high c-p values [25, 26].

Despite essential differences between these types of response of the nematode community after a disturbance, both response types can be seen as a retrogression in the succession of the nematode community; that is, from the perspective of the r- and K-concept and in terms of the c-p classification. Therefore, the MI is not only an index to monitor natural succession, but can also be applied to detect disturbances and to monitor any subsequent recovery of the ecosystem. A review of applications of the MI in environmental studies comprising among others effects of water pollution, heavy metals, eutrophication, oil spill, liming, acidification, physical stresses and tillage regimes is given by De Goede & Bongers, [17]. These studies showed that in general, disturbances are followed by a decrease in MI, whereas recovery or natural succession coincided with an increase in MI.

Bioindicators should give opportunities to distinguish between different stress factors [27]. De Goede *et al.* [23] indicated that the MI-concept offers possibilities to discriminate between effects of eutrophication and pollution, which correspond to changes in the composition of the nematode fauna following the earlier mentioned response types I and II, respectively. Examples of studies demonstrating response type I involve effects of eutrophication, fertilization, oil spill, liming and several physical stresses [17]. These disturbances cause such an increase in nutrient availability that, as a result of the (temporal) increased microbial activity, c-p 1 taxa ('enrichment opportunists' *sensu* Ettema & Bongers, [24]) can increase in number. When the nematode food supply decreases the densities of the c-p 1 taxa will decrease, while often at the same time the densities of c-p 2 taxa can increase. The higher scaled taxa (c-p 3-5) hardly react numerically to these changes in food supply [24].

At present, examples of the type II response are less often documented. Type II responses were found for soil acidification [25] and for the chronic effects of heavy metals [26, 28]. Although further confirmation is required, it seems that a type II response can be expected for long-term effects of disturbances where toxic stress is accompanied with conditions of low food availability. Under such conditions most taxa, but especially those belonging to higher c-p groups, decrease in number and the food availability is to low to have high populations of c-p 1 taxa. Certain taxa from the c-p 2 group (the 'general opportunists' *sensu* Ettema & Bongers, [24]) are the most tolerant to survive these severe conditions, or even benefit of it and increase in number [26].

The response types I and II both result in a decrease of the MI. As a rule of thumb a decrease of the MI below 2 as a consequence of an absolute increase found among c-p 1 taxa, usually indicates a type I response (eutrophication), whereas samples characterized by MI values close to 2, and a lowered species diversity consisting mainly of taxa with c-p value 2, indicates a type II response (pollution). However, it is possible that in habitats with pollution, organic inputs can lead to a period in which c-p 1 taxa predominate the nematode fauna such that the MI is below 2. Under these conditions the MI is less effective to identify pollution (type II response).

To reduce this effect of enrichment in measuring pollution induced stress, it might have advantages to omit taxa scaled in c-p 1 from the calculation of the MI. We propose to express this new index as the MI2-5. Compared to the MI, the MI2-5 has proved to be more sensitive to indicate pollution as demonstrated in Table 1.. Moreover, comparison of the scores of both indices can help to discriminate between pollution induced stress or eutrophication. First applications of this MI2-5 can be found in Figure 1., [29] and in [30].

TABLE 1. Correlation coefficients (r) for heavy metal content versus MI and MI2-5 for 20 agricultural soils in The Netherlands with mixed pollutants [31]

	MI	MI2-5
Copper	-0.300	-0.496
Zinc	-0.318	-0.502
Cadmium	-0.178	-0.466
Lead	-0.154	-0.461

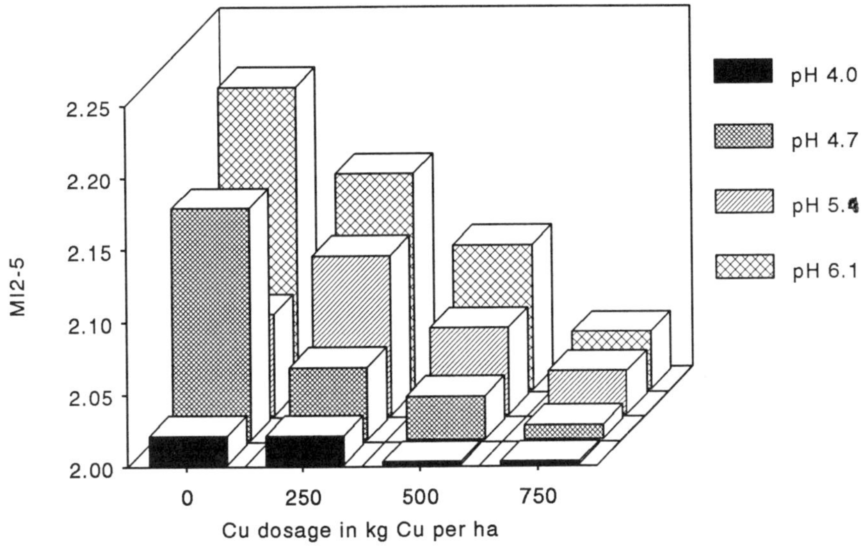

Figure 1. Effect of Cu and pH on the nematode MI2-5.

5. Discussion

The classification of nematode taxa into c-p groups reflects a first attempt to categorize nematodes in ecological groups with similar life-history characteristics. Although promising results were obtained when using this concept, further calibration of the c-p scaling might be necessary. Moreover, substantial more knowledge is needed on the relationship between the life-history of certain species and their sensitivity towards soil pollution. Figure 2., for example, shows the effects of Cu on two closely related genera within the Cephalobidae. Although both genera have the same c-p value (c-p 2), their sensitivity towards Cu differs greatly. When it becomes possible to unravel these relationships, any improvements of c-p scaling can result in an increased sensitivity of the MI and MI2-5.

In addition to the MI and MI2-5, examination of the changes within each of the c-p groups on which the indices are based, can provide additional information on the nature of the underlying changes in the habitat. De Goede *et al.* [18] presented a c-p triangle in which they schematically indicated the main directions along which changes in the nematode fauna composition may occur. Since this method makes it possible to visualize the changes in c-p 1, 2 and 3-5 distribution, it can also be used to identify type I and type II responses within the nematode community. Moreover, examination of c-p groups can also be helpful to detect simultaneous increase or decrease of opportunists and persisters which may remain unnoticed when using only the MI.

As was mentioned in the recommendations of the NATO Advanced Research Workshop (Moscow, 1995) ecological relevant bioindicator systems should, among others, be based on information obtained from different species and processes. Studies carried out at the community level provide opportunities in this direction. However, in order to interpret complex changes in the structure of whole communities, there is an urgent need to develop new methods which can be used by decision makers as well. The MI-concept can become one of these tools, especially because the changes in MI not only indicate the disturbance acting on a nematode community, but also provide opportunities to identify specific types of disturbance. Therefore, it seems worthwhile to investigate the usefulness of this MI-concept also for other groups of organisms, especially for those organisms that, in comparison to nematodes, have a different route of exposure.

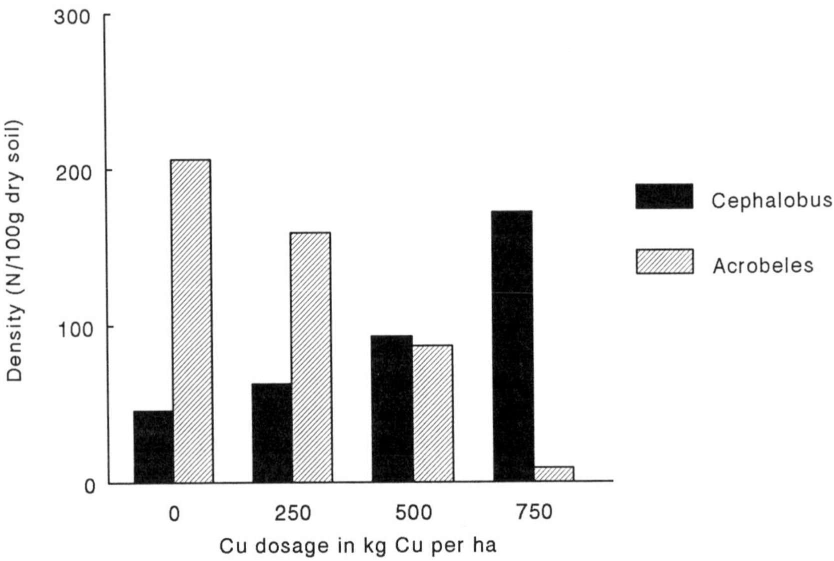

Figure 2. Effect of Cu on the density of *Cephalobus* and *Acrobeles* at a nominal pH-KCl of 6.1.

References

1. Moriarty, F. (1983) *Ecotoxicology. The study of pollutants in ecosystems*, Academic Press, London.
2. Cairns, J. (1992) The threshold problem in ecotoxicology, *Ecotoxicology* **1**, 3-16.
3. Moore, J.C. and Ruiter, P.C. de (1993) Assessment of disturbance on soil ecosystems, *Veterinary Parasitology* **48**, 75-85.
4. Clements, W.H. (1994) Assessing contaminant effects at higher levels of biological organization, *Environ. Toxicol. Chem.* **13**, 357-359.
5. Bongers, T. (1990) The maturity index: an ecological measure of environmental disturbance based on nematode species composition, *Oecologia* **83**, 14-19.
6. Odum, E.P. (1969) The strategy of ecosystem development, *Science* **164**, 262-270.

7. MacArthur, R.H. and Wilson, E.O. (1967) *The theory of island biogeography*, Princeton University press, Princeton.
8. Pianka, E.R. (1978) *Evolutionary ecology*, Harper & Row, New York.
9. Brown, V. and Southwood, T.R.E. (1987) Secondary succession: patterns and strategies, in A.J. Gray, M.J. Crawley and P.J. Edwards (eds.), *Colonization, succession and stability*, Blackwell, Oxford, pp. 315-337.
10. Wilbur, H.M., Tinkle, D.W. and Collins, J.P. (1974) Environmental certainty, trophic level, and resource availability in life history evolution, *Am. Nat.* **108**, 805-817.
11. Parry, G.D. (1981) The meanings of r- and K-selection, *Oecologia* **48**, 260-264.
12. Southwood, T.R.E. (1977) Habitat, the template for ecological strategies, *J. Anim. Ecol.* **46**, 337-365.
13. Nichols, J.D., Conley, W., Batt, B. and Tipton, A.R. (1976) Temporally dynamic reproductive strategies and the concept of r- and K-selection, *Am. Nat.* **110**, 995-1005.
14. Bengtsson, J. and Baur, B. (1993) Do pioneers have r-selected traits? Life history patterns among colonizing terrestrial gastropods, *Oecologia* **94**, 17-22.
15. Bongers, T., Alkemade, R. and Yeates, G.W. (1991) Interpretation of disturbance-induced maturity decrease in marine nematode assemblages by means of the maturity index, *Mar. Ecol. Progress Series* **76**, 135-142.
16. Bongers, T., Goede, R.G.M. de, Korthals, G.W., and Yeates, G.W. (1995) Proposed changes of c-p classification for nematodes, *Russian J. of Nematol.* **3**, 61-62.
17. Goede, R.G.M. de and Bongers, T. (1995) The nematode Maturity Index. Manual of methods for research on the ecology of nematodes, *Techniques in nematode ecology*, Soc. of Nematologists, in press.
18. Goede, R.G.M. de, Georgieva, S.S., Verschoor, B.C. and Kamerman, J.W. (1993) Changes in nematode community structure in a primary succession of blown-out areas in a drift sand landscape, *Fundamental and applied Nematology* **16**, 501-513.
19. Goede, R.G.M. de, unpubl. results.
20. Regier, H.A. and Cowell, E.B. (1972) Applications of ecosystem theory, succession, diversity, stability, stress, and conservation, *Biol. conserv.* **4**, 83-88.
21. Whittaker, R.H. (1975) Communities and ecosystems, MacMillan, New York.
22. Odum, E.P. (1985) Trends expected in stressed ecosystems, *BioScience* **35**, 419-422.
23. Goede, R.G.M., Bongers, T. and Ettema, C. (1993) Graphical presentation and interpretation of nematode community structure: C-P triangles, *Meded. Fac. Landbouwwet. Univers. Gent* **58**: 743-750.
24. Ettema, C.H. and Bongers, T. (1993) Characterization of nematode colonization and succession in disturbed soil using the Maturity Index, *Biol. and Fertil. of Soils* **16**, 79-85.
25. Ruess, L., Funke, W. and Breunig, A. (1993) Influence of experimental acidification on nematodes, bacteria and fungi: soil microcosms and field experiments, *Zool. Jb. Syst.* **120**, 189-199.
26. Korthals, G.W., Alexiev, A.D., Lexmond, Th.M., Kammenga, J.E. and Bongers, T. (1995) Long-term effects of copper and pH on the nematode community of an agroecosystem, *Envir. Tox. and Chem.*:in press.
27. Bongers, T. and Schouten, T. (1991) Nematodengemeenschappen als potentieel diagnostisch instrument voor chemische verontreinigingen, in G.P. Hekstra and F.J.M. van Linden (eds.), *Flora en fauna chemisch onder druk*, Pudoc, Wageningen, pp. 175-186.
28. Popovici, I. and Korthals, G.W. (1995) Soil nematodes used in the detection of habitat disturbance due to industrial pollution, *Stud. Univ. Babes-Bolyai, Biol.*, **XXXVIII**, 1993, 1-2.
29. Korthals, G.W., Bongers, T., Alexiev, A.D., and Lexmond Th.M. (1993) Influence of copper and pH on a terrestrial nematode community, Poster presented at SETAC World Conference, March 1993, Portugal.
30. Popovici, I. (1994) Nematodes as indicators of ecosystem disturbance due to pollution, *Stud. Univ. Babes-Bolyai, Biol.* **37(2)**, 1992, 15-27.
31. Bongers, T. (1992) De nematodenfauna als bio-indicator voor de bodemkwaliteit, internal report.

THE USE OF MACRO-INVERTEBRATES FOR POPULATION AND COMMUNITY MONITORING OF METAL CONTAMINATION - INDICATOR TAXA, EFFECT PARAMETERS AND THE NEED FOR A SOIL INVERTEBRATE PREDICTION AND CLASSIFICATION SCHEME (SIVPACS)

D. J. SPURGEON, R. D. SANDIFER and S. P. HOPKIN
University of Reading
Ecotoxicology Research Group
School of Animal and Microbial Sciences
P.O. Box 228, Reading, RG6 6AJ
United Kingdom

Abstract

The loss of species from a system will lead to changes in ecosystem structure, energetics and processes. Consequently, the changes in abundance and in particular the presence or absence of species from areas where they would normally be expected to occur should be the final measure of the biological impact of a pollutant. If such studies are to be useful for monitoring pollutant impact, the techniques should be simple and relatively quick to undertake. The number of taxa that can be incorporated into such studies may be limited to reduce time demands. Suitable indicator taxa should conform as closely as possible to the seven criteria outlined by Pearson which are described in this paper. Three taxa of terrestrial macro-invertebrates that appear to fit these criteria are oligochaeta, aranaea and isopods. Surveys of macro-invertebrate community structure were undertaken at sites contaminated with cadmium, copper, lead and zinc, situated around a smelting works. Results indicate that community sensitivity can vary considerably between taxa. Earthworms were relatively sensitive to metals, whilst no deleterious effect on the total catch, number of species or diversity of spiders was detected. As a consequence of the differential sensitivities existing between taxa, community resilience for any given group should not necessarily be regarded as evidence of ecosystem health. A broader based scheme developed with acknowledgement to RIVPACS offers a more suitable basis for assessing pollutant induced changes at the community level.

N. M. van Straalen and D. A. Krivolutsky (eds.), Bioindicator Systems for Soil Pollution, 95–110.

1. Why monitor at the community level?

The role of biomonitoring is to indicate the "health" of ecosystems in which it is suspected that a biologically significant concentration of a pollutant has been released. As early as the 1950s, it was suggested that biological systems should provide the basis for environmental assessments [1, 2], as the extrapolation of effects from abiotic samples is confounded by a variety of biotic and non-biotic factors.

Hopkin [3] outlined four biological approaches for monitoring the effects of pollutants in ecosystems: 1) Assessing impact on community structure, 2) Measuring pollutant concentrations in the tissues of sentinel species, 3) Quantifying effects on individual performance, and 4) Detecting the presence of genetically altered races, resistant to pollutants. All these strategies have a role within any monitoring framework, for example, strategy 2 is useful for assessing the transfer of pollutants to higher organisms [3], strategy 2 for appraisal of effects on individual species such as beneficial arthropods [4] and strategy 4 for pesticide management [5].

It is our opinion that assessments of the health of ecosystems at contaminated and remediated sites should be made by determining changes in community structure at such sites. In particular, the absence of species from areas where they would normally be expected to occur should be considered a strong indication that deleterious effects have occurred. Changes in abundance or the removal of species will lead to changes in system energetics and ultimately, variations in systems processes and food webs [6]. Since the impact of pollutants on ecosystems will be derived from their cumulative effects on constituent species, the protection of community structure should ensure that the structure and function of the receiving ecosystem are preserved.

2. Why monitor with invertebrates?

There are a number of valid reasons for using invertebrate species for community monitoring. Major invertebrate groups are present over a range of diverse habitats, are relatively easy to sample and can be simple to identify when good keys are available [7]. In addition, the low mobility of some groups such as earthworms means they are representative of the habitat being sampled. This is not always the case for groups such as vertebrates.

Disadvantages of using terrestrial invertebrates include problems in species level identification, since the taxonomy of some groups is not clear [8], and limitations of the sampling techniques [9]. Additionally, sample processing and identification can be time consuming. Thus, the rapid enumeration of results from multiple site comparisons is impeded.

The implication of time constraints on the development of community monitoring will be discussed later in this chapter.

3. The development of community monitoring in aquatic and terrestrial systems

Community monitoring with macro-invertebrates has been widely used in aquatic systems. The earliest schemes were based on the clearly identifiable community responses that occur after the organic enrichment of freshwater. Sites were scored for contamination based on the presence or absence of species with presumed sensitivity to organic pollution [8, 10]. Although useful for assessing organic enrichment, score systems were not applicable for assessing the impact of a range of pollutants, since relative taxal sensitivities may vary.

The second generation of schemes concentrated on measuring the richness and equilibrity of distribution of individuals between species resulting in the development of diversity indices such as Shannon-Wiener's and Simpson's. Although these indices are still widely used for monitoring purposes, their usefulness is somewhat restricted, since conclusions from the various systems do not always correlate at the same sites and perhaps more seriously, it is not always clear that diversity decreases with increasing pollution [3]. For example, Bengtsson and Rundgren [11, 12] found a bell-shaped response of diversity for spiders, harvestmen, slugs, beetles, ants and springtails along a copper and zinc gradient with maximum diversity at intermediate concentrations.

The most recently developed community assessment scheme in aquatic systems is the River Invertebrate Prediction and Classification System (RIVPACS). This system uses a multi-variant model to predict the fauna of an unperturbed site of known character. Actual communities can then be compared to these targets allowing potential biotic effects of pollutants to be assessed [13]. This system allows predictions of relative abundances, biotic scores and species and family occurrence to be made. The system was developed in the UK and is now in routine used by regulatory authorities. The scheme has attracted interest from countries in both Europe and the rest of the world.

Community monitoring is at an earlier stage of development in terrestrial systems. It has long been recognised that contaminants can alter community structure in terrestrial ecosystems. For example, metal rich spoil tips of disused mines often support a distinct flora of tolerant or pre-adapted metallophytes [14], while alterations in the relative abundance of Collembola species have been found in contaminated regions [12, 15]. However, no strategy is currently in widespread use in Europe for monitoring the impact of chemicals using terrestrial invertebrates. The success of such schemes in aquatic systems suggests the need to develop an assessment procedure applicable to the terrestrial environment.

For some groups of pollutants the most suitable communities for assessing environmental impact will be defined by the nature of the chemical under study. For example, the non-target effects of pesticides may be best monitored by measuring their effects on populations of selected beneficial invertebrates. In such circumstances the best techniques may be adapted from population biology. However, for pollutants with more general effects, simple population measurements are unlikely to be useful in determining effects, since it will not always be clear which species will be most sensitive for a given chemical and thus what should be monitored. For these pollutants, a more integrated system of monitoring is required.

4. Developing community monitoring in terrestrial ecosystems

Community monitoring studies with invertebrates have the potential to provide a widely applicable means of assessing the impact of pollutants. However, as outlined in Section 1, a number of problems exist in using invertebrates as monitoring organisms. As previously stated, a major limitation is the amount of time required for sampling, particularly if multiple sites are used to define response gradients. For this reason, previous studies have tended to focus on assessing effects on specific invertebrate groups [11, 12, 16, 17, 18]. The reduced workload required for such studies is likely to appeal to the regulatory agencies, as these organisations will demand monitoring schemes that are quick to undertake. However, the implications of limiting monitoring to a single group on the conclusions made from such studies have not been examined. Clearly, results from community studies could differ profoundly depending on what is monitored.

Pearson [19] has suggested that faunal groups suitable for community monitoring should correspond as closely as possible to the following criteria: 1) Taxonomy must be well known and stable, 2) Natural history must be well known, 3) They must be readily surveyable, easy to manipulate and identify, 4) Higher taxa (order, family, tribe and genus) must have a broad geographical distribution over a breadth of habitat types, 5) Lower taxa (species and subspecies) must be specialised and sensitive to habitat changes, 6) Taxa must have potential economic importance, 7) Patterns of biodiversity must be reflected in other related and unrelated taxa. Three taxa of terrestrial invertebrates that appear to fit these criteria well are earthworms, woodlice and spiders.

In Table 1 each of the three groups has been scored for each of Pearson's criteria to a maximum of five. No score for the relationships of sensitivity between taxa has been awarded, since it is not clear if the concept of such relationships fits current ecotoxicological philosophies where ideas of a single representative "sensitive" species have largely been rejected. In general all three groups score well. The taxonomy of these groups can be considered reasonably stable (compared for example to springtails and mites) and they are well represented among the published keys [20, 21, 22]. The natural histories are well known [23, 24, 25, 26, 27, 28], and there are a number of standardised techniques existing for their survey. Higher taxa have a broad range of distribution, while particularly for spiders, species are sensitive to habitat differences. Earthworms are of economic importance as they maintain soil fertility, while spiders are important predators of pests.

All three taxa fit Pearson's criteria well and could be used within the scope of a community monitoring assessment. However, it is pertinent to ask how the choice of group would affect the conclusions made from a community study at a polluted site? To examine the relationships between the results of monitoring studies with these three taxonomic groups, results of studies with earthworms, woodlice and spiders along a pollutant gradient are compared. To assess the community responses of each taxon, the groups were surveyed at sites in the region around a zinc, lead, cadmium smelting works situated at Avonmouth, in the south-west of England. This region is contaminated with high concentrations of cadmium, copper, lead and zinc derived from emissions from the smelter.

TABLE 1. The suitability of three invertebrate taxa for community monitoring as described by their correspondence to the seven characteristics for an assessment group detailed by Pearson [19] Groups are scored to a maximum of five for each property except for "relations to other taxa", as it was considered that insufficient data was available to score for this characteristic (see text)

Criterion	Earth-worms	Isopods	Spiders
Taxonomic stability	3	4	4
Knowledge of natural history	5	5	4
Surveyability	3	5	4
Distribution of higher taxa	5	4	5
Habitat change sensitivity of lower taxa	3	3	5
Economic importance	5	2	4
Sensitivity relations to other taxa	?	?	?
Total	24	23	26

At sites closest to the smelter, metal concentrations of 300 μg Cd g^{-1}, 3000 μg Cu g^{-1}, 15,000 μg Pb g^{-1} and 35,000 μg Zn g^{-1} have been recorded [29] and grid based soil sampling has indicated an exponential decline in metal concentrations with distance from the factory (see [30, 31, 32] for a full discussion of the spatial distribution of metals in the Avonmouth area).

The aim of this study was to determine the community responses of these groups to the high soil metal concentrations to ascertain the interrelationship of results from monitoring studies with different groups. For each taxon, a number of community based parameters were considered. These were total abundance (total catch), total number of species, mean number of individuals per species, Shannon-Wiener diversity, total biomass and mean individual biomass (see [33] for details of parameters).

5. Details of the study

Results from two separate studies are given. For the first study, the populations of earthworms at 22 sites in the region around the smelter were assessed by collection from four 0.25 m x 0.25 m quadrats at each site. All earthworms were weighed and identified using the key of Sims and Gerrard [22]. In the second study, spider and woodlouse populations were measured at five sites during July 1994, the peak period of abundance for spiders. At each site, six pitfall traps were laid. These were then left for one month after which they were pooled and all spiders and woodlice that had entered the traps identified to groups, family and ultimately, where possible, species. The biomass of woodlice and spiders was not determined.

A number of papers have been published criticising the use of pitfall traps for the assessment of ground active invertebrate populations [9]. Catches from pitfall traps can be influenced by factors including species, sex, hunger level, pattern and spacing of traps, size, climate and season. There is also a growing volume of evidence suggesting that certain species such as carabids and linyphiid spiders can actively avoid traps and thus catches are reduced [34, 35]. Sunderland *et al.* [9] concluded that the correct term to describe the catch of pitfall traps should be "activity-trappability-density" to alert authors that trapping efficiency varies between species and habitat. However, in this study, "total catch" will be used.

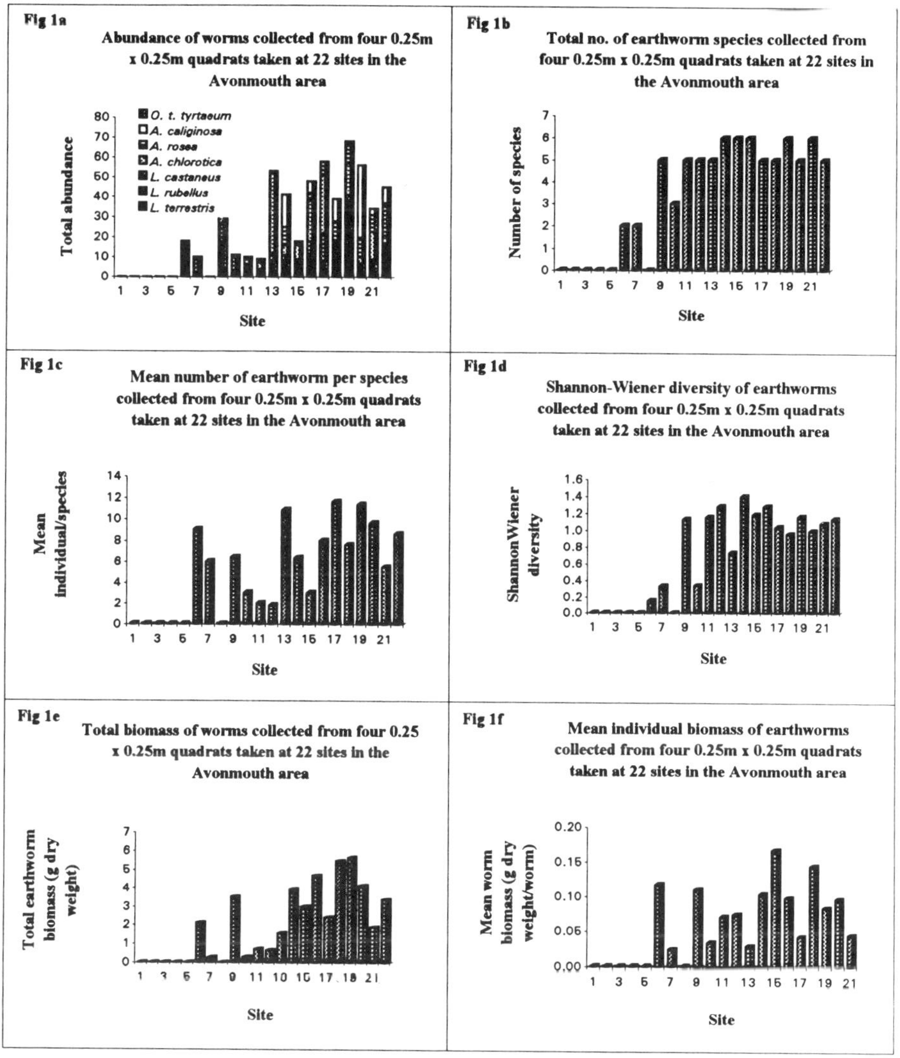

Figure 1. Changes in earthworm communities as indicated by effects on six parameters at 21 sites in the region around a smelting works situated at Avonmouth in south-west England (Sites 1-21) and an unpolluted control site (Site 22). Site 1 is situated within 0.5 km from the smelting works, Site 21 is 10 km from the factory, while Site 22 is over 100 km distant from the pollution source. Six different community parameters are given in the figures a to f.

6. Community response to a gradient of metal contamination

6.1. EARTHWORMS

Earthworm abundance was reduced at sites up to 3 km from the factory. Of particular importance was the absence of all earthworms from the five closest sites (Fig. la). Lack of worms is almost certainly responsible for the considerable accumulation of undecomposed leaf litter at site close to the smelter [31]. If the abundance data for the individual species are analysed, it is clear that there are differences in sensitivity Three species, *Aporrectodea rosea, Allolobophora chlorotica* and *Aporrectodea caliginosa* were more sensitive to metals, since these were absent from sites close to the smelter where *Lumbricus rubellus, Lumbricus castaneus* and *Lumbricus terrestris* persisted (Fig. 1a).

Differences in sensitivity between earthworms resulted in a decrease in the number of species collected from an individual site (Fig. lb). However, the absence of species from the fauna was not accompanied by increasing the abundance of more resilient worms. As a consequence, the mean number of individuals per species was not altered at any of the sites from which worms were collected (Fig. lc). Shannon-Wiener diversity showed a marked decrease with increasing soil metal concentrations. Lowest diversities were recorded at the sites nearest to the factory where the least number of species were present (Fig. 1d).

Total earthworm biomass decreased with distance from the smelting works, reflecting the decrease in earthworm abundance at sites close to the smelter (Fig. le). Mean individual biomass did not show any clear trend (Fig. lf). Thus, there is no evidence for a change in the size structure of the earthworm populations at the most contaminated sites.

Abundance, total number of species, diversity and biomass all showed a logarithmic pattern of decline with increasing proximity to the smelter. As a consequence, it is possible to fit logistic models to the data to calculate EC_{50} values for each of these parameters. For these calculations, the toxic unit (TU) model of Sprague [36] has been applied to determine a value based on the assumption of the additive effects of metals. Although as stated by Spurgeon *et al.* [37] and Spurgeon and Hopkin [29] the most important toxic effect for earthworm populations at sites close to the smelter is due to the effects of zinc. Calculation of EC_{50} values for abundance, total number of species, diversity and biomass gave values of 2.45, 7.35, 4.6 and 1.75 TU g^{-1} respectively. Values could not be calculated for number of individuals per species and mean individual biomass, since these parameters did not decrease sigmoidally with increasing metal concentration. It is noticeable, that the lowest values are for the parameters that show greatest inter-site variation (biomass and abundance). This inter-site variability gives a shallow slope for the logistic regression and thus lower EC_{50} values are calculated.

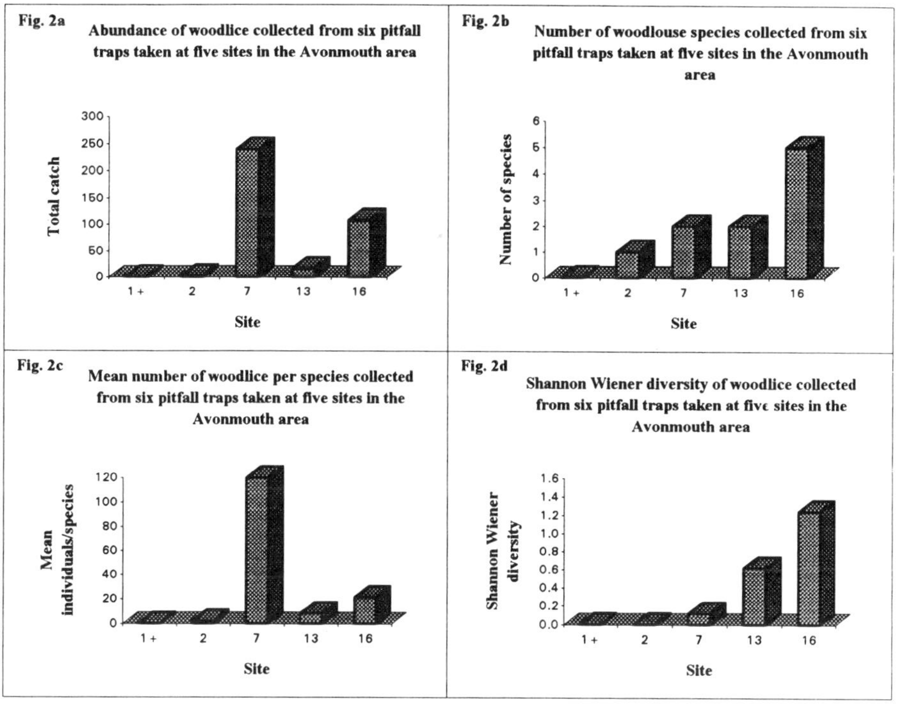

Figure 2. Changes in isopod communities at five sites in the region around a smelting works situated at Avonmouth, south-west England, as indicated by effects on four measured parameters (figures a to d). Sites are numbered with reference to the previous earthworm community study (see legend to Figure 1). Site 1+ is located closer to the smelter than any site previously used.

6.2. WOODLICE

A clear trend for isopod catches to decrease was not found, due primarily to the low catches at Site 13. The differences in total catches at the less contaminated sites indicate that there is high inter-site variability. However, a reduction in catches at the most contaminated site was recorded, with all isopods absent from the two most contaminated sites (Fig. 2a). The number of isopod species was clearly reduced by increasing metal concentrations in soil. Three isopod species, *Philoscia muscorum, Trichoniscus pusillus* and *Androniscus dentiger* were present at Site 16, but were not recorded at any of the four remaining sites. At these sites only *Porcellio scaber* and *Oniscus asellus* were present. Despite the reduction in the number of species, no consistent pattern in reduction of the number of individuals per species was found at any sites where isopods were present (Fig. 2c).

Isopod diversity was lower at sites close to the smelter (Fig. 2d). Because of the low number of sites used, it is not possible to fit logistic curves for the calculation of EC_{50} values. Consequently, it is not possible to state, which parameter is most sensitive for assessing the impact of metals on isopod communities. However, from the data, it is clear that species number and diversity are among the most sensitive parameters, while number of individuals per species is least sensitive.

6.3. SPIDERS

Surprisingly perhaps, the total catches of spiders tended to increase at sites close to the factory (Fig. 3a). In particular, number of lycosid spiders of the *Pardosa* genus increased greatly. Both increases in the actual abundance of spiders and raised activity of individuals could account for the increase in total catches. The high concentrations of metals present at Site 1+ (over 5 % of the soil dry weight) result in a sparse vegetation cover in areas of the site. The presence of bare soil patches is known to favour ground hunting spiders such as lycosids, which like to bask and hunt in the open. However, there was also a reduction in the catches of some prey groups in the Site 1 pitfalls. For example, springtail and mite catches were reduced, although the numbers of ants and particularly orthopterans are increased (Sandifer, unpublished data). The decreased abundance of some prey may demand that spiders hunt over a larger area to increase their chances of capture.

No clear reduction in the total number of species was found at the most contaminated sites, indeed, like total catch, the number of species tended to increase with increasing metal concentrations (Fig. 3b). The number of individual per species found at the sites closest to the factory was also increased at the most contaminated sites, due primarily to the high catches of lycosids (Fig. 3c). Spider diversity was slightly reduced at the most contaminated site (Fig. 3d). This decrease can be related to the increased lack of equilibrity resulting from the increase in lycosid numbers.

7. Implications of the results for the development of community monitoring systems

Results from community studies with the three chosen taxa indicate the range of sensitivity existing between groups in their responses to soil pollutants. As a result of these differences, the conclusions drawn from studies with individual groups can vary considerably, depending on the taxa choosen for monitoring. In the Avonmouth area, sampling of spider

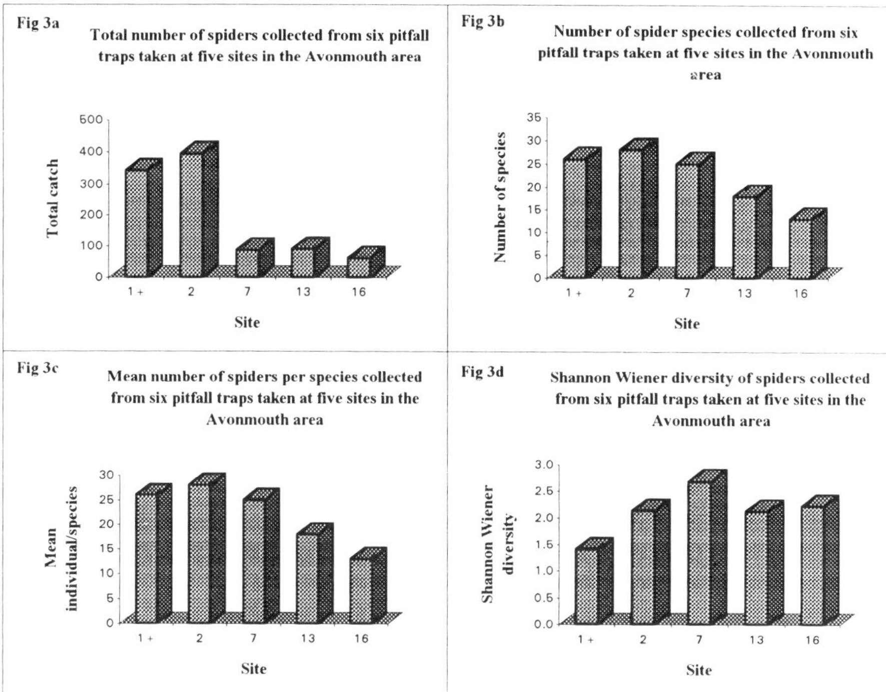

Figure 3. Changes in spider communities at five sites in the region around a smelting works situated at Avonmouth, south-west England, as indicated by effects on four measured parameters (figures a to d). Sites are numbered with reference to the previous earthworm community study (see legend to Figure 1). Site 1+ is located closer to the smelter than any site previously used.

communities would not indicate any gross effects due to metals even at the most contaminated site, while the earthworm community shows effects at sites up to 3 km from the smelter.

Of the three groups studied, earthworm communities appear most suitable for determining the impact of metals, since this taxon shows clear dose-related responses along the pollution gradient. Thus, from this study it could be concluded that earthworms would be an ideal group for monitoring the effects of pollutants at the community level. However, considerable evidence exists to suggest that the relative sensitivities of taxa vary considerably between pollutants. For example, spiders are relatively sensitive to a number of pesticides to which earthworms are comparatively tolerant, whilst, as has been shown, the reverse is true for metals. These differences in relative sensitivity between groups present a fundamental problem for monitoring with individual taxa. As stated earlier, the search for a "most sensitive species" for laboratory toxicity testing has largely been abandoned because it was concluded that the most sensitive species would vary considerably for each pollutant [38]. The search for a most sensitive indicator taxon for field monitoring may be expected to befall such a fate.

A further problem in community monitoring is that results will be flexible in time as well as space [33, 39]. The development of resistance in invertebrates to the effects of pollutants is a widely recognised phenomenon. Over 500 cases of resistance have been reported among arthropods for pesticides alone [40], whilst a number of species have been shown to develop both physiological and life history adaptations under metal stress [39]. The development of resistance in invertebrate groups may occur at different rates. As a consequence, some species will be able to colonise the most polluted regions of a gradient more rapidly. Knowledge of the development of resistance in different species such as that detailed by Donker and Bogert [41] will be important for the correct interpretation of community study results.

The most sensitive parameter for community assessment also varies between groups. Lowest EC_{50} values for earthworms were obtained for biomass and abundance, although these were also the parameters with the greatest inter-site variation (Figs. 1a-f). For woodlice, the species number and diversity appeared most useful for assessing community structure changes, whilst for spiders, only diversity indicated any deleterious effects at the most contaminated sites. For the latter two groups, measurement of total numbers collected (abundance), appears insufficient to indicate deleterious impact on communities. In these groups, it is essential that individuals are identified to species level to allow calculations of diversity and total species number, if deleterious effects on communities are to be detected.

8. Conclusions

A classic ecotoxicological problem exists that may hinder the development of a successful community monitoring strategy for assessing the effects of pollutants on ecosystems. Regulatory agencies are likely to demand that studies are quick to undertake (and thus have a limited scope), while the demand for successful prognosis of effects means that it is important to undertake surveys of the maximum number of groups, at the maximum possible taxonomic resolution. How can these two conflicting demands be satisfied to develop a non pollutant-specific community biomonitoring scheme?

The results detailed in this paper indicate that it is important to include as many major groups as possible in any survey, as profound effects can occur for a given group at pollutant concentrations at which other groups appear unaffected. The inclusion of a broad range of groups in a study will result in a large workload if all individuals are to be identified to species. However, if in the initial survey, taxonomic resolution, particularly of the most difficult groups is reduced, this may decrease the workload, while ensuring as broad a scope as possible Altematively, species level identification could be focused on groups for which it appears that effects on abundance can be detected at the lowest concentration.

Perhaps the most satisfactory system for assessing contaminated sites would be to predict the presence of a number of key species based on the physiological characteristics of the site. Actual fauna could then be assessed to determine if these species are present. The groups where key species are absent could then be subjected to more intense examination to determine trends in the pattern of species removal across the site. This system could be then used to determine the the extent of contamination at a given site and could allow site classification across a broad range of pollutants. Such a "Soil Invertebrate Prediction and Classification System (SIVPACS)", developed with reference to RIVPACS, a system already existing for monitoring the effects of pollutants in freshwater ecosystems (see Section 3), would be a great advance for assessing the impact of soil contaminants at the ecosystem level.

9. References

1. Patrick, R. (1951) A proposed biological measure of stream conditions, *Verh. Internat. Verein. Limnol.* **11**, 299-307.
2. Hynes, H.B.N. (1960) *The Biology of Polluted Waters,* Liverpool Univ. Press, Liverpool.
3. Hopkin, S.P. (1993) *In situ* biological monitoring of pollution in terrestial and aquatic ecosystems, in P. Calow (ed.), *Handbook of Ecotoxicology, Volume 1,* Blackwell, Oxford, UK, pp. 397-427.
4. Jepson, P.C. (1993). Insects, spiders and mites, in P. Calow (ed.), *Handbook of Ecotoxicology, Volume 1,* Blackwell, Oxford, UK, pp. 299-325.
5. Denholm, I., Devonshire, A.L. and Hollomon, D.W. (1992) *Resistance '91. Achievements and Developments in Combating Pesticide Resistance*, Elsevier Applied Science, London, UK.
6. Moore, J.C., De Ruiter, P.C., and Hunt, H.W. (1993) Soil invertebrate/micro-invertebrate interactions: disproportionate effects of species on food web structure and function, *Vet. Parisitol.* **48**, 247-260.

7. Sims, R.W., Freeman, P., and Hawksworth, D.L. (1988) *Key Works to the Fauna and Flora of the British Isles and North-western Europe,* Clarendon for the Systematics Association, Oxford, UK.
8. Reynoldson, T.B. and Metcalfe-Smith, J.L. (1992) An overview of the assessment of aquatic ecosystem health using benthic invertebrates, *J. Aquat. Ecosys. Health* **1**, 295-308.
9. Sunderland, K.D., De Snoo, G.R., Dinter, A., Hance, T., Helinius, J., Jepson, P., Kromp, B., Samu, F., Sotherton, N.W., Ulber, B., and Vangsgaard, C. (1993) Density estimation for beneficial predators in agroecosystems, *Acta Jutlandica* **88,** 1-17.
10. Metcalfe, J.L. (1989) Biological water quality assessment of running waters based on macro-invertebrate communities: History and present status in Europe, *Environ. Pollut.* **60**, 101-139.
11. Bengtsson, G. and Rundgren, S. (1984) Ground-living invertebrates in metal-polluted forest soils, *Ambio* **13**, 29-33.
12. Bengtsson, G. and Rundgren, S. (1988) The Gusum case: a brass mill and the distribution of soil Collembola, *Can. J. Zool.* **66**, 1518-1526.
13. Wright, J.F., Armitage, P.D., and Furse, M.T. (1989) Prediction of invertebrate communities using stream measurements, *Reg. Rivers Res. Manage.* **4**, 147-155.
14. Baker, A.J.M. and Proctor, J (1990) The influence of cadmium, copper, lead and zinc on the distribution and evolution of metallophytes in the British Isles, *Plant System. Evolut.* **69**, 89-104.
15. Hågvar, S. and Abrahamsen, G. (1990) Microarthropoda and Enchytraeidae (Oligochaeta) in a naturally lead contaminated soil: a gradient study, *Environ. Entomol.* **19**, 1263-1277.
16. Bengtsson, G. and Rundgren, S. (1982) Population density and species number of enchytraeids in coniferous forest soils polluted by a brass mill, *Pedobiologia* **24**, 211-218.
17. Bengtsson, G., Nordström, S., and Rundgren, S. (1983) Population density and tissue metal concentration of Lumbricids in forest soils near a brass mill, *Environ. Pollut. (Ser. A.)* **30**, 87-108.
18. Read, H.J., Wheater, C.P., and Martin, M.H. (1987) Aspects of the ecology of Carabidae (Coleoptera) from woodlands polluted by heavy metals, *Environ. Pollut.* **48**, 61-76.
19. Pearson, D.L. (1994) Selecting indicator taxa for quantitative assessment of biodiversity, *Phil. Trans. R. Soc. Lond.* **B 345**, 75-79.
20. Hopkin, S.P. (1991) A key to the woodlice of Britain and Ireland, *Field Studies* **7**, 599-650.

21. Roberts, M.J. (1993) *The Spiders of Great Britain and Ireland,* Harley Books, Colchester, Essex, UK.
22. Sims, R.W. and Gerrard, B.M. (1985), in D.M. Kermack and R.S.K. Barnes (eds.), *Earthworms. Synopsis of the British Fauna (New Series)*, Linnean society, London, UK.
23. Edwards, C.A. and Lofty, J.R. (1977) *Biology of Earthworms,* 2nd Edition, Chapman and Hall, London.
24. Ferrara, F. (ed.) (1989) Proceedings of the Second Symposium on the Biology of Terrestrial Isopods, *Monitore Zool. Ital. (New Series)* **4**.
25. Lee, K.E. (1985) *Earthworms: Their Ecology and Relationship with Soil and Land Use,* Academic Press, London, England.
26. Foelix, R.F. (1982) *The Biology of Spiders,* Harvard University Press, Cambridge, Massachusetts, USA.
27. Sutton, S.L. (1972) *Woodlice,* Pergamon Press, Oxford.
28. Sutton, S.L. and Holdich, D.M. (1984) *The Biology of Terrestrial Isopods,* Oxford Science Publications, Clarendon Press, Oxford, UK.
29. Spurgeon, D.J. and Hopkin, S.P. (in press). Extrapolation of the laboratory-based OECD earthworm test to metal-contaminated field sites. *Ecotoxicology*
30. Jones, D.T. (1991) *Biological Monitoring of Metal Polluton in Terrestrial Ecosystems,* Ph.D. Thesis, University of Reading, UK.
31. Hopkin, S.P (1989). *Ecophysiology of Metals in Terrestrial Invertebrates,* Elsevier Applied Science, London, UK.
32. Martin, M.H. and Bullock, R.J. (1994) The impact and fate of heavy metals in an oak woodland ecosystem, in S.M. Ross (ed.), *Toxic Metals in Soil-plant Systems*, John Wiley, Chichester, UK, pp. 327-365.
33. Gray, J.S., Aschan, M., Carr, M.R., Clarke, M.R., Clarke, K.R., Green, R.H., Pearson, T.H., Rosenberg, R., and Warwick, R.M. (1988). Analysis of community attributes of the benthic macrofauna of Frierfjord/Langesundfjord and in a mesocosm experiment, *Mar. Ecol. Prog. Ser.* **46**, 151-165.
34. Benest, G. (1989) The behaviour of a carabid when facing the trap, *Rev. Écol. Biol. Sol* **26**, 505-514.
35. Topping, C.J. (1993) Behavioural responses of three linyphiid spiders to pitfall traps, *Entomologia Exp. Appl.* **68**, 287-293.
36. Sprague, J.B. (1970) Measurement of pollutant toxicity to fish. II. Utilizing and applying bioassay results, *Water Res.* **4**, 3-32.
37. Spurgeon, D.J., Hopkin, S.P. and Jones, D.T. (1994) Effects of cadmium, copper, lead and zinc on growth, reproduction and survival of the earthworm *Eisenia fetida (Savigny):* Assessing the

environmental impact of point-source metal contamination in terrestrial ecosystems, *Environ. Pollut.* **84,** 123-130.

38. Cairns, Jr, J. (1986) The myth of the most sensitive species, *Bioscience* **36**, 670-672.
39. Posthuma, L. and Van Straalen, N.M. (1993) Heavy-metal adaptation in terrestrial invertebrates: A review of occurrence, genetics, physiology and ecological consequences, *Comp. Biochem. Physiol.* **106C,** 11-38.
40. Klerks, P.L. and Levington, J.S. (1993) Evolution of resistance and changes in community composition in metal-polluted environments: a case study of Foundry Cove, in R. Dallinger and P. S. Rainbow (eds.), *Ecotoxicology of Metals in Invertebrates,* Lewis, Chelsea, USA, pp. 223-240.
41. Donker, M.H. and Bogert, C.G. (1991). Adaptation to cadmium in three populations of the isopod *Porcellio scaber, Comp. Biochem. Physiol.* **100C**, 143-146.

THE PROBLEM OF SCALE IN BIOINDICATION OF SOIL CONTAMINATION

A.D. POKARZHEVSKII
Institute of Evolutionary Animal Morphology and Ecology
Russian Academy of Sciences
Leninsky Prospekt 33
117071 Moscow
Russian Federation

Abstract

Soil bioindication studies have often shown that species of different size groups show reactions to contaminants that were not expected on the basis of laboratory data. This phenomenon may be explained by the hierarchical organization of soil ecosystems in which different scales of time and space are involved. The lowest level is represented by ecosystems of unicellular and tiny multicellular organisms inhabiting or living in association with water films in soil pores and litter cavities. At the intermediate level there are ecosystems of small multicellular organisms dwelling in soil pores and litter cavities as a whole. Large multicellular organisms use the soil as a whole and they live in the soil ecosystem of the highest level. We prefer the latter organisms as bioindicators because they live in the same time-space scale as we. An early warning system for ecotoxicological effects can, however, not be elaborated on the basis of a single size group. A bioindicator system for soil pollution has to include organisms from differently-sized ecosystems and so take the problem of scale into account.

1. Introduction

A major difficulty associated with the interpretation of any bioindication or ecotoxicological study is that differently-sized species show different responses to toxicant exposure, in experiments and in the field. In addition, the reaction to toxicant exposure in the field may be quite different from the reaction shown in laboratory experiments [1, 2, 3]. These problems have to be solved before ecotoxicological risk

N. M. van Straalen and D. A. Krivolutsky (eds.), Bioindicator Systems for Soil Pollution, 111–121.

assessment methods based on soil organism studies can be applied successfully [4, 5].

The mechanisms underlying the ecotoxicity of soil pollutants have been discussed previously [5, 6, 7]. One of the main factors is the sensitivity of a species to a toxicant. A second factor is the physiological state of the species observed, while a third important factor is the time-space dimension of the ecosystem in which the species lives, including the interactions with other organisms. This paper will concentrate on problems of scale in time and space, and the consequences of scale for the population responses towards the emission of toxicants into the soil environment. I will argue that it is necessary to include spatial and temporal dimensions in any risk assessment system that aims to predict the real risk of soil contamination.

2. Size relationships in soil

Ghilarov [8] has shown that there is a clear negative correlation between the abundance of soil animals and the body size of the species. Edwards [9] underlined that there are different biotic relationships between soil animals of different size groups. Ceitlin and Byzova [10] observed that different size groups of soil animals have a similar biomass, on a logarithmic scale. It can thus be supposed that species of the same size group form a network of trophic and other interactions on their own, relatively independent of other size groups. Tight nutrient cycles have to be maintained within such systems [11], while interactions with other systems take place by means of organic matter fluxes and oxygen produced by higher plants and algae. The systems of small dimensions develop inside the systems of larger size.

Ghilarov [12] has also underlined that the soil environment provides a habitat for three groups of soil organisms: water film dwellers (protozoans, rotifers, tardigrades), soil pore dwellers (microarthropods and other microfauna species), and real soil dwellers (earthworms and macroarthropods). Each group exploits a food resource independent from the other groups: the first group feeds on bacteria and algae, or animals of the same size, the second group feeds on fungi, or animals of the same size, and the third one feeds on plant tissue (living or dead), or animals of the same size. During its development, an individual animal can, however, move from one group to another. For example, the young stages of the lithobiid *Monotarsobius curtipes* feed on springtails, since they are in the same size group, but the adult animals do not use springtails anymore [13]. The same phenomenon is known for spiders.

Large soil animals consume the ecosystems of smaller dimensions as a whole, and then use them as an essential part of their metabolism. Earthworms, diplopods, isopods, larvae and imagoes of insects consume soil, litter and plant roots including the bacteria-algae-Protozoa ecosystems developed on them. These ecosystems undergo further development in the guts of earthworms or macroarthropods and supply saprophageous animals with essential amino acids, phosphorus, and vitamins [14, 15]. Soil and litter can be considered as sources of substrates, organic matter and nutrients for the ecosystems present in animal guts. In termites or ruminants, the gut ecosystems have become completely independent from the external environment and are connected with it only through the materials consumed by the animal.

In view of the considerations given above, one may distinguish in soil three relatively closed and independent ecosystems with different dimensions in space-time. The first ecosystem is the **bacteria-algae-Protozoa** ecosystem, consisting of unicellular organisms as well as some multicellular organisms such as tardigrades, rotifers, bacterivorous mites, springtails, and nematodes. These ecosystems have a size of a few mm to some cm, and a volume of some cubic cm. They are isolated in water fims of soil cavities or at root and litter surfaces. The ecological time of these ecosystems (the time of developing through one successional sequence) is in the order of days to months; biological turnover time (the time it takes for nutrient fluxes to replace nutrient stores) varies from a day to a week. During succession, the trophic structure of the system develops and populations of organisms at higher trophic levels appear. The predator component of this ecosystem is equivalent to the functional group in Moore *et al.* [16]

The second ecosystem is the **fungi-microarthropod** ecosystem, consisting of small multicellular organisms, including mites, springtails, nematodes, enchytraeids, and juvenile stages of large soil animals. These ecosystems are isolated in soil pores and cavities in the litter layer. They have a size varying from some cm to a few m, and are often limited by the rhizosphere volume of separate plants. Ecological time varies from a week to some months, while biological turnover time is in the order of days to months. The league concept of Faber [17] is the consumer component of this ecosystem.

The third ecosystem is the **earthworm-plant** ecosystem, consisting of large multicellular organisms. Its borders coincide with those of a plant community above ground. Ecological time varies from some years to decades, while biological turnover takes some months or more, up to some years.

An artist impression of the three ecosystems, including information on size and time relationships, is given in Fig. 1. It is important to realize that the ecosystems of lower levels are nested in the ecosystems of higher levels. A mosaic or patchy distribution of Protozoa and microarthropods is a reflection of the spatial distribution of small-scale ecosystems, and does not relate to the patchiness of the large-scale ecosystem in which it is nested.

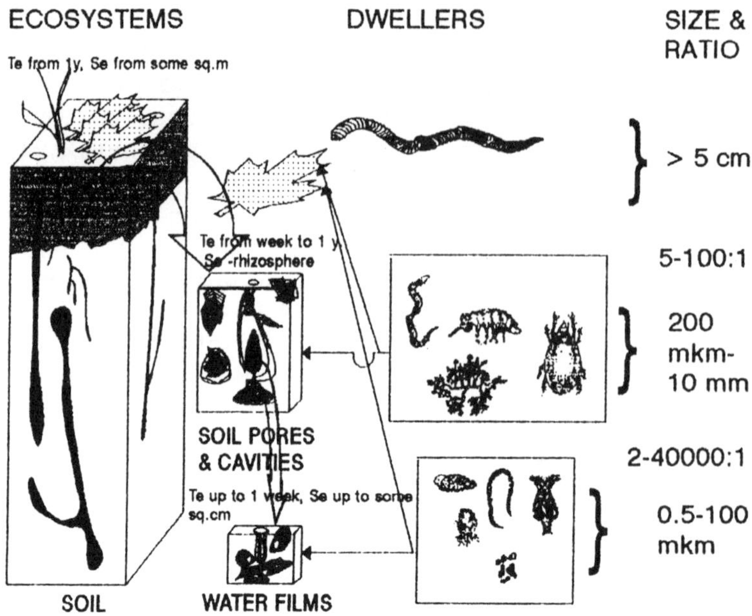

Figure 1. The proposed hierarchical concept of soil ecosystems, showing how the bacteria-algae-Protozoa ecosystem is nested in the fungi-microarthropod ecosystem, which in turn is nested in the earthworm-plant ecosystem. Te = ecological time (relating to succession), Se = ecosystem size.

Is it legitimate to consider these systems of different dimensions as separate ecosystems? They are connected with each other by means of energy fluxes and oxygen produced by plants and algae (although plants and algae belong to different systems). In the sense of Tansley [18] the three systems described here are to be considered as real ecosystems. The term "subsystem" is not convenient in this context, because Swift and Anderson [19] have used it already to describe the trophic structure of ecosystems; they distinguished "plant subsystem", "grazing subsystem", and "decomposer subsystem".

3. Consequences for bioindication and ecotoxicology

In a recent review, Eijsackers [6] has discussed the problem of scale in the context of the ecotoxicology of soil organisms. He paid attention to the fluctuations in environmental conditions, which have different consequences for organisms of different sizes. We may ask the question, however, is it really necessary to define ecosystems of different scales? Are there any qualitative differences between the ecological processes in ecosystems of different scales? Some examples from bioindication and radioecological studies demonstrate that this is indeed the case.

In a field study on the influence of ^{90}Sr on soil Protozoa [1] it was found that populations of Protozoa, which are the most tolerant organisms towards ionizing radiation, were affected to a greater extent than populations of microarthropods, which are actually more sensitive [3, 20]. This is demonstrated in Fig. 2, showing data on soil animal populations affected by radioactive contamination in Southern Ural. In another study, conducted in the surroundings of the Chernobyl nuclear power plant after the accident in 1986 [2], microarthropod populations responded more obvious to radioactive contamination than earthworm populations, see Fig. 3. Similar responses were revealed in field studies with experimentally manipulated radioactive radiation [21]. It appears that radiation doses in small spaces (water films, soil pores) are much higher than expected on the basis of the measured (bulk) radioactivity of a soil.

The presence or absence of a single species of a tiny animal has a neglible effect on the soil as a whole, however, changes in mesofauna communities following application of pesticides may lead to changes in the decomposition of plant residues [22, 23]. Differences in the species composition of large animals may lead to marked changes in soil and in biogeochemical processes [24, 25]. The appearance or disappearance of microarthropods usually remains unnoticed, except for some parasites or

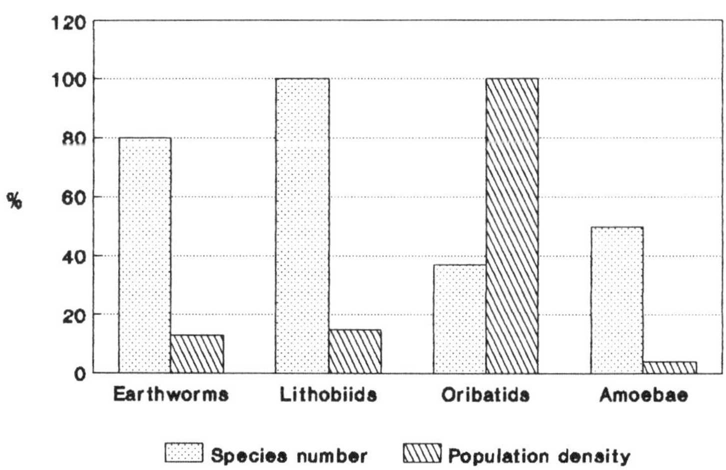

Figure 2. Changes (in % of the initial level) in populations of soil invertebrates of different size groups, affected by ^{90}Sr contamination in Southern Ural [1, 3, 20].

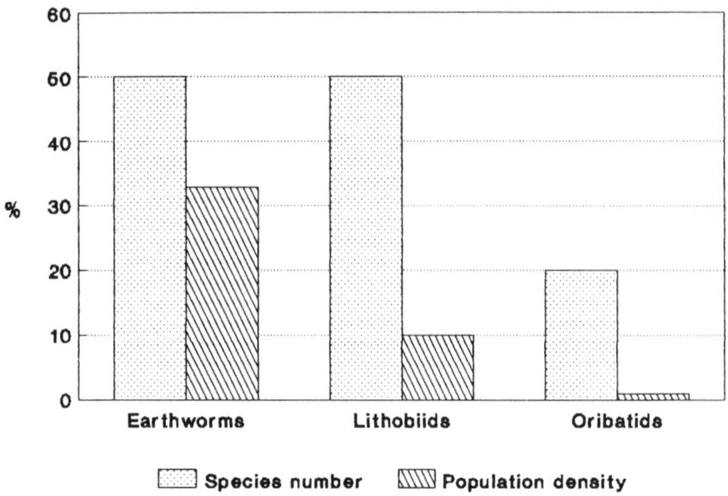

Figure 3. Changes (in % of the initial level) in populations of soil invertebrates of different size groups, affected by radioactive contamination at Chernobyl [2].

hosts of animal or plant diseases. Consequently, the role of a large animal in the biological turnover of a community is equivalent to the role of a group of small animals [11]; in ecosystems of their own dimensions, however, small species play the same role as large species in their ecosystems. As a consequence of the size-selective effects of soil ecosystems, the greatest number of soil species is known for the size group 5-10 mm, and the number of species declines sharply for smaller or larger groups [26].

An important consideration when interpreting bioindication data is that small animal species can specialize on particular food species. Different species of springtails were able to select a particular microfungus to feed upon, and the preference was influenced by heavy metal contamination of the substrate on which the fungi were growing [27, 28]. It is obvious that protozoans, tardigrades and rotifers, within the dimensions of their ecosystems, must be capable of selecting prey items that we cannot identify as separate species.

One of the reasons why earthworms are absent from plots contaminated by heavy metals is that contamination alters the composition of the microbial community [29]. In other words, the disappearance of an ecosystem at a low level may lead to the disappearance of an ecosystem at the highest level.

4. Bioindication on the basis of the hierarchical system

Sukachev, in his "biogeocoenosis" [30] proposes a hierarchical system of ecosystems of at least three size levels (not in the sense of O'Neill *et al.* [31]). These systems are more or less closed, connected to each other only through fluxes of organic matter, without trophic interactions between them. What is the relevance in elaborating a bioindication system for the assessment of soil pollution and other human influences? Some consequences of the considerations given above may be listed as follows.

1. The influence of the same factor in soil ecosystems of different dimensions is qualitatively different. An ecosystem of a certain scale cannot not be used as a model for ecosystems of other dimensions, and the extrapolation of data between systems of different scales may be incorrect. This has to be taken into account in any bioindication study.

2. Organisms that live in an ecosystem of the first level will reflect toxicant impacts already on a short term. This is a result of the short life-cycles of most of the organisms, a high rate of pollution-induced natural selection in the populations, and a quick succession, resulting in a relatively short exposure time for an individual ecosystem. Another reason why small organisms reflect toxicant impacts on a short term is related to the high degree of patchiness of the toxicant's distribution at the microscale level. In addition, small organisms take up toxicants mainly from the water phase, where it is present in a soluble and in highly bioavailable form.

3. Organisms inhabiting ecosystems of the second scale level reflect the mid-term impacts of a toxicant. This is the result of a relatively long life-cycle of most of the species (compared to the organisms in the first level), a lower genetic flexibility of the populations, and a relatively long exposure time of the ecosystems. There are physico-chemical barriers, such as organic matter and microorganisms developing on it, that hamper the uptake of toxicants by organisms.

4. Large organisms are indicators of long-term effects of toxicants in soil, due to long life-cycles, a role in soil forming processes, and an influence on soil organisms living in lower level ecosystems. In turn, effects on ecosystems of lower scale will often influence organisms at the higher scale later.

Considering the above listed arguments, we have to develop a bioindication and ecotoxicological system that includes organisms from soil ecosystems of different dimensions. A conceptual scheme for such a system is proposed in Fig. 4. For practical purposes, concrete organisms and procedures have to be selected. Many of the organisms and procedures have been considered before, and the selection of methods for field bioindication does not seem to present many difficulties [3, 32]. A coherent set of organisms for risk assessment and early warning of the

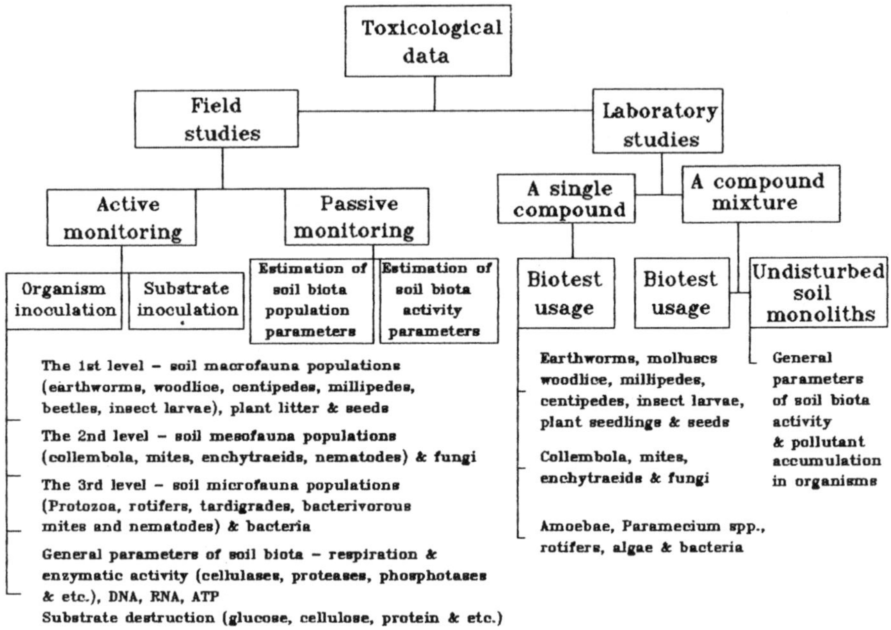

Figure 4. A conceptual scheme for a soil bioindication system

ecotoxicity of different compounds and their mixtures is required. At the moment, only *Eisenia foetida* as the model organism and decomposition as the model proces are being considered for the assessment of ecotoxicity of new compounds [33].

To develop risk assessment methods for soil contamination, there is a continuing search for new organisms and new procedures [34, 35], however, most of them are for large animals. In my opinion, it is more important to elaborate standard procedures for reference organisms of the lower level ecosystems. Enchytraeids, springtails, nematodes and rotifers deserve attention due to ease of cultivation [34, 35, 36] and ease of exposure methods. There is also a method for predicting effects of toxicants in mixtures, the "undisturbed monolith method", that is used for the assessment of toxicant mixtures [37, 38].

Acknowledgements. This work was granted by the Federal Programme "Ecological Safety of Russia", theme 5.1.4.3. I thank Prof. Dmitrii Krivolutskii for discussions on the problem and for valuable remarks.

5. References

1. Korganova, G.A. (1973) An influence of experimental soil contamination by Strontium-90 on soil Protozoa, *Zool. J.* **52**, 939-941 (in Russian).
2. Krivolutskii, D.A. and Pokarzhevskii, A.D. (1992) Effects of radioactive fall-out on soil animal populations in the 30 km zone of the Chernobyl atomic power station accident, *Sci. Total Environ.* **112**, 69-77.
3. Krivolutskii, D.A. (1994) *Soil Fauna in Ecological Control,* Nauka, Moscow (in Russian).
4. Van Straalen N.M. and Denneman, C.A.J. (1989) Ecotoxicological evaluation of soil quality criteria, *Ecotox. Environ. Safety* **18**, 241-251.
5. Van Straalen, N.M., Leeuwangh, P., and Stortelder, P.B.M. (1994) Progressing limits for soil ecotoxicological risk assessment, in M.H. Donker, H. Eijsackers, and F. Heimbach (eds.), *Ecotoxicology of Soil Organisms,* CRC Press Inc., Boca Raton, pp. 397-409.
6. Eijsackers, H. (1994) Ecotoxicology of soil organisms: seeking the way in a pitch-dark labyrinth, in M.H. Donker, H. Eijsackers, and F. Heimbach (eds.), *Ecotoxicology of Soil Organisms*, CRC Press Inc., Boca Raton, pp. 3-32.
7. Løkke, H. (1994) Ecotoxicological extrapolation: tool or toy, in M.H. Donker, H. Eijsackers, and F. Heimbach (eds.), *Ecotoxicology of Soil Organisms*, CRC Press Inc., Boca Raton, pp. 411-425.
8. Ghilarov, M.S. (1994) Size and number relationships of soil invertebrates, *Reports of the USSR Acad. Sci.* **43**, 283-285 (in Russian).
9. Edwards, C.A. (1967) Relationships between weights, volumes and numbers of soil animals, in O. Graff and J.E. Satchell (eds.), *Progress in Soil Zoology*, Vieweg & Sohn, Braunschweig, pp. 585-594.
10. Ceitlin, V.B. and Byzova, J.B. (1986) Size distribution of soil organisms in different nature zones, *J. Obschei Biologii* **47**, 193-197 (in Russian).
11. Pokarzhevskii, A.D. (1987) Species specificity in biogenic turnover, in R. Volkis (ed.), *The Status of Species among Biological Systems*, Inst. Zool. Parasitol. Lithuan SSR, Vilnus, pp. 90-97 (in Russian).
12. Ghilarov, M.S. (1949) *Features of Soil as a Habitat and its Significance in Insect Evolution*, The USSR Academy of Sciences, Moscow and Leningrad (in Russian).

13. Sergeeva, T.K. (1983) Zoophagy of *Monotarsobius curtipes* (Chilopoda, Lithobiomorpha) in Moscow outskirt spruce forests, *Zool. J.* **62**, 1003-1008 (in Russian).
14. Pokarzhevskii, A.D., Sikora, I., and Gordienko, S.A. (1984) Amino acid resources in saprophage feeding, *Reports of the USSR Academ. Sci.* **277**, 253-256 (in Russian).
15. Pokarzevskii, A.D. (1990) Bioindicational and ecotoxicological aspects of investigation into the trophic structure of animal communities, in D.A. Krivolutsky (ed.), *Bioindication of Chemical and Radioactive Pollution*, Mir Publ., Moscow.
16. Moore, J.C., Walter, D.E., and Hunt, H.W. (1988) Arthropod regulation of micro- and mesobiota in below-ground detrital food-webs, *Ann. Rev. Entomol.* **33**, 419-439.
17. Faber, J.H. (1992) Functional classification of soil fauna: a new approach, *Oikos* **62**, 110-117.
18. Tansley, A.G. (1935) The use and abuse of vegetational concepts and terms, *Ecology* **16**, 284-307.
19. Swift, M.J. and Anderson, J.M. (1989) Decomposition, in H. Leith and M.J.A. Werger (eds.), *Tropical Rain Forest Ecosystems, Biogeographical and Ecological Studies*. Ecosystems of the World, No. 14B, Elsevier, Amsterdam, pp. 547-569.
20. Krivolutskii, D.A. (1983) *Radioecology of Terrestrial Animal Communities*, Energoatomizdat, Moscow (in Russian).
21. Krivolutskii, D.A., Tikhomirov, F.A., Federov, E.A., Pokarzhevskii, A.D., and Taskaev, A.I. (1988) *Influence of Ionizing Radiation on Biogeocoenosis*, Nauka, Mosow (in Russian).
22. Seastedt, T.R. (1984) The role of microarthropods in decomposition and mineralization processes, *Ann. Rev. Entomol.* **29**, 25-46.
23. Pokarzhevskii, A.D. (1985) *Geochemical Ecology of Terrestrial Animals*, Nauka, Moscow (in Russian)
24. Ghilarov, M.S. (1965) *Zoological Method of Soil Biodiagnostics*, Nauka, Moscow (in Russian).
25. Edwards, C.A., Reichle, D.E., and Crossley, D.A. Jr. (1970) The role of soil invertebrates in turnover of organic matter and nutrients, in D.E. Reichle (ed.), *Analysis of Temparate Forest Ecosystems*, Ecol. Studies 1, Springer Verlag, Berlin, pp. 147-172.
26. Peters, R.H. (1983) *The Ecological Implications of Body Size*, Cambridge University Press, Cambridge.
27. Bengtsson, G., Gunnarson, T., and Rundgren, S. (1983) Growth changes caused by metal uptake in a population of *Onychiurus*

armatus (Collembola) feeding on metal polluted fungi, *Oikos* **40**, 216-225.
28. Bengtsson, G., Ohlson, L., and Rundgren, S. (1985) Influence of fungi on growth and survival of *Onychiurus armatus* (Collembola) in a metal polluted soil, *Oecologia (Berl.)* **68**, 63-68.
29. Bengtsson, G., Gunnarsson, T., and Rundgren, S. (1986) Effects of metal pollution on the earthworm *Dendrobaena rubida* (Sav.) in acidified soils, *Water Air Soil Pollut.* **28**, 361-383.
30. Sukachev, V. and Dylis, N. (eds.) (1964) *Fundamentals of Forest Biocoenology*, Nauka, Moscow (in Russian).
31. O'Neill, R.V., DeAngelis, D.L., Waide, J.B., and Allen, T.F.N. (1986) *A Hierarchical Concept of Ecosystems*, Princeton University Press, Princeton.
32. Donker, M.H., Eijsackers, H., and Heimbach, F. (eds.) (1994) *Ecotoxicology of Soil Organisms*, CRC Press Inc., Boca Raton.
33. *Guidelines for Testing Chemicals* (1984) Organization for Economic Co-operation and Development (OECD), Paris.
34. Eijsackers, H. and Løkke, H. (1992) *SERAS, Soil Ecotoxicological Risk Assessment System*, Report National Environmental Research Institute, Silkeborg, Denmark.
35. Løkke, H. and Van Gestel, C.A.M. (1993) Manual of SECOFASE, Report National Environmental Research Institute, Silkeborg, Denmark.
36. Westheide, W. and Bethke-Beilfuß, D. (1991) The sublethal enchytraeid test system: guidelines and some results, in G. Esser and D. Overdieck (eds.), *Modern Ecology: Basic and Applied Aspects*, Elsevier, Amsterdam, pp. 497-508.
37. Terytze, K., Mentscher, E., Pokarzhevskii, A., Gusev, A., Zhulidov, A., and Savtschenko, L. (1989) Methodischer Ansatz zur Einschatzung der Kombinationswirkungen von Substanzen in der Umwelt, *Nachrichten Mensch-Umwelt* **H1**, 44-61.
38. Pokarzhevskii, A.D., Martjushov, V.Z., Tarasov, O.V., and Antonova, T.A. (1993) Quantitative parameters of radioisotope mixture effects on soil biological activity, in V.E. Sekolov and D.A. Krivolutskii (eds.), *Ecological After-effects of the Radioactive Contamination at South Ural*, Nauka, Moscow.

BIODIVERSITY INDICATORS FOR SUSTAINABILITY. ASSESSMENT OF RURAL LANDSCAPES

M.G. PAOLETTI and D. SOMMAGGIO
Biology Department, Padova University
Via Trieste 75, 35100 PADOVA

Abstract
Biological organisms are tha base of life on our planet. Following recent evaluations, only 1/20 to 1/60 of our planet's species have yet been described and most of these will be lost if the destruction rate of the environment continues at the actual rate.
Most of the terrestrial environment (up to 95%), is affected by human activities including agriculture and the terrestrial habitats provide up to 98% of human food on the planet. Sustainable strategies of food production in agriculture have to improve the existing biodiversity and assist in the following: increased porosity of the landscape through proper management of natural vegetation, better use and recycling of organic residues, introduction of integrated farming systems, reduced tillage, rotation, biological control, increased number of biota involved in human foodwebs. Attention must be paid in western countries to non conventional food, insects and other terrestrial invertebrates, such as microlivestock.
In addition, small organisms such as soil invertebrates can offer interesting opportunities to monitor the landscape as a suitable tool to evaluate environmental degradation and pollution and sustainability in general. The species number, abundance, dominance, and foodweb complexity, can offer a basic tool to evaluate the rural landscape and it situation. Examples chosen from our research carried out in Italy include apple orchards, peach orchards and vineyards.

N. M. van Straalen and D. A. Krivolutsky (eds.), Bioindicator Systems for Soil Pollution, 123–140.

Invertebrates such as carabids, isopods, earthworms, seem, for instance, suitable to give sufficiently detailed information.

1. Introduction

The plurality of living organisms on the planet, together with energy and nutrients, form the base of our living food chains. The use of resources for human needs have varied and transformed patterns of biological diversity in the landscape. Rural areas and human settlements in general dominate terrestrial environments and some considerations on the existing situation are required, for instance, use of species, to revise the needs for a sustainable future. Reflections on landscape structure and functions to maintain biodiversity and its use seem pressing (Paoletti, 1995). A collection of data and modelling of biodiversity in non disturbed as well in rural landscapes is needed as part of the transition of agriculture from a conventional, high input model, towards a sustainable ones both ecologically and economically. We have experienced a set of approaches using invertebrates as tool to assess sustainability of existing farming systems directly based on farm field work and using different invertebrate as basic indicator tools.

2. Data base of biodiversity: a missing point.

Living species, are the base of the planet's biological diversity. This living core of organisms is the base for all human activities (Wilson, 1988; Pimentel et al., 1992).

At the moment no exhaustive data base on living species exists. For this reason, estimations of existing described species oscillate between 1.3 million (Wilson, 1988; Wheeler, 1990) and 1.8 million (Stork, 1988). In Fig 1 the distribution of the described living organisms is represented by specimens covering in proportional areas. Note the large area occupied by Insects.

However, the forecast species are some orders of magnitude even more abundant on the planet. Erwin (1982) first documented the incredible projection of insects, using the canopy of some South American rainforests as a model; he suggested the existence of over 30 million species (May, 1992). More recently Ehrlich and Wilson (1991) have estimated that living species could reach the 100 million mark! In fact in the last few years "the fondness of God for beetles and insects in general" has been extended to many other taxa such as bacteria, fungi and many small invertebrates like mites and nematodes (Paoletti et al., 1992).

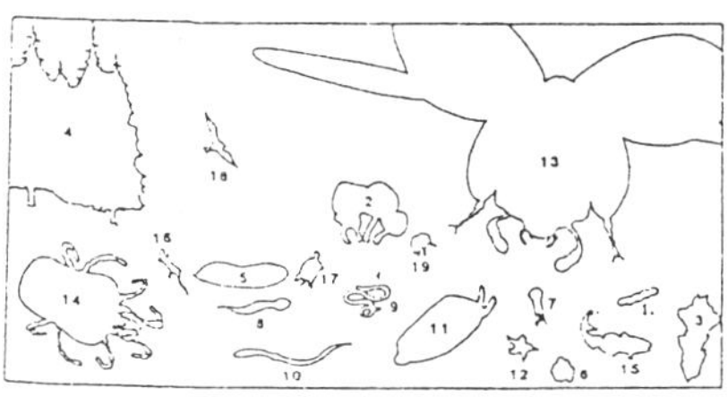

	Taxon	No. of described species
(1)	Monera (bactreria, blue-green algae)	4760
(2)	Fungi	46983
(3)	Algae	26900
(4)	Plantae (multicellular plants)	248428
(5)	Protozoa	30800
(6)	Porifera (sponges)	5000
(7)	Coelenterata (jelly fish, corals, comb jellies)	9000
(8)	Platyhelminthes (flatworms)	12200
(9)	Nematoda (roundworms)	12000
(10)	Anellida (earthworms, etc.)	12000
(11)	Mollusca (molluscs)	50000
(12)	Echinodermata (starfish, etc.)	6100
(13)	Insecta	751000
(14)	Non insect Arthropoda (mites, spiders, crustaceans, etc.)	123161
(15)	Pisces (fish)	19056
(16)	Amphibia (amphibians)	4184
(17)	Reptilia (reptiles)	6300
(18)	Aves (birds)	9040
(19)	Mammalia (mammals)	4000

Figure 1. Number of living species described (Wheeler, 1990). The size of different taxa are designed in relation to the described species. □: 1000 described species (from Wheeler, 1990).

3. Biodiversity as a base for human food chains.

There are at least two points which amaze the researcher; how many beetles and insect species live on the planet and how few plant and animal species we currently consider as our possible food. In Western countries, for instance, insects as well as most small invertebrates are still considered inedible, in spite of the evidence supporting insects to be the large majority of living organisms.

However, over 500 species of insects are eaten world-wide, especially in tropical and Far Eastern countries (DeFolliart, 1989, 1990). A approximately 90% of world food for people comes from just 15 plant and 8 animal species (Wilson, 1988), however, the use of biodiversity is incredibly different among different human groups. In Java small farmers cultivate 607 crop species in their gardens, with an overall species diversity comparable to deciduous subtropical forests (Dover and Talbot, 1987; Michon, 1983). In Swaziland, Africa, 220 wild plant species are commonly consumed (Ogle and Grivetti, 1985). Andean farmers cultivate many clones of potatoes, more than 1000 of which have names (Clawson, 1985). The traditional dish pistic in the Friuli region Italy it sum up to 56 different wild herbs (Paoletti et al., 1995a).

4. Rearing the Forest

The story of rural landscapes in Western countries is in many ways associated with the progressive destruction of forests. To stress the abundance of species in the forest it is sufficient to note the recent figures given by Buszko (1991) for a small area of Polish deciduous woodlands near Torun, in which 1211 species of Lepidoptera have been recorded! Forests, compared with other environments, suggest a higher biodiversity (Table 1). In spite of the fact that no extensive research has been developed in comparing "natural" versus "cultivated" environments (Paoletti et al., 1992), the data available seem to sponsor higher diversity of invertebrates in forested areas especially in the tropics (Hansen et al., 1991). However if natural vegetation or hedgerows and woodland remnants are part of the rural landscape, diversity is increasing (Pimentel et al. 1992; Paoletti et al., 1989).

TABLE 1. Animal biological diversity in ecosystems.

Ecosystem	Location and area	Animal species	Source
Pasture	Britain (ha)	1000	McFadyen, 1961
Tropical forest	Borneo (500 ha)	6200[a]	Hammond, 1991
Tropical forest	Borneo (10 trees)	2800[b]	Stork, 1988
Tropical forest	Costa Rica (10800 ha)	13000[b]	Janzen, 1987
Beech forest	Germany (forest)	1500-1800	Ellenberg et al., 1986
Deciduous forest	Poland (37 ha)	1211[c]	Buszko, 1991
Hungarian National Parks (mostly forest)	Hungary	4433-8843d	Mahunka, 1987
Collards	New Yorks (ha)	262	Root, 1973
Alfalfa	New York (ha)	600[e]	Pimentel and Wheeler, 1973
Corn and woodland remnants	NE Italy	450[f]	Paoletti, 1988
Corn agroecosystems and deciduous woodland remains (projections)	NE Italy	800-1500[e,f]	Paoletti, 1988
Sustainable farming systems, Lautenbach: mesostigmatic mite	Germany	60	El Titi and Landes, 1989
Hungarian corn field	Hungary	598[e]	Meszaros, 1984a
Hungarian apple orchard	Hungary	1759[d]	Meszaros, 1984b

[a]Beetles only.
[b]Insects only.
[c]Lepidoptera only.
[d]All animals.
[e]Aboveground arthropods.
[f]Soil invertebrates.
Data from: Paoletti and Pimentel, 1992; Paoletti et al., 1992.

5. Biodiversity in Agroecosystems

Sustainable agricultural systems could sustain a higher number of species in rural landscapes. A higher sensibility towards ecological problems and the economic problem associated with conventional agriculture (soil erosion, ground water pollution, high input cost, etc.) helped in developing more sustainable agriculture with reduced external input. Table 2 is an attempt to incorporate some trends at farm and landscape scales that can be effective in preserving biodiversity and economically sustainable farming systems. Nevertheless, at a regional level appropriate research to compare undisturbed ecosystems and agroecosystems in a historical perspective is also needed. In fact few data on biodiversity are available for many regions. Clearly, to preserve biodiversity on the planet, increased understanding and modeling of managed agricultural landscape versus forest ecosystems are needed.

TABLE 2. Farming systems which can increase biodiversity in agroecosystems (from Paoletti et al., 1992)

Sustained biodiversity	Decreased biodiversity
Hedgerows (11,12)	Wild vegetation removal
Dikes with wild herbage (11,12)	Tubular drainage or removal
Polyculture (1)	Monoculture
Agroforestry (13)	Monoculture
Rotation with legumes (10)	Monosuccession
Dead mulch, living mulch (7,10)	Bare soil
Strip crops, ribbon cropping (*)	Conventional cropping
Alley cropping (*)	Monoculture
Minimum, no tillage, ridge tillage (7,16,17)	Conventional ploughing
Mosaic landscape structure porosity (8,9,15)	Landscape simplification woodland clearance
Organic sustainable farming (5,10)	Intensive input farming
On-farm research (13,14)	Conventional plot research
Organic fertiliser (5,10)	Chemical fertiliser
Biological pest control (6)	Conventional chemical pest control
Plant resistance (6)	Plant susceptibility
Germplasm diversity (1,2)	Standardisation

References: 1, Altieri et al., 1987; 2. Lal, 1989; 3, Oldfield and Alcorn, 1987; 5, Matthey et al., 1990; 6, Pimentel et al., 1991; 7, Stinner and House, 1990; 8, Paoletti, 1988; *, personal evaluation (Paoletti, 1987-1990); 9, Noss, 1987; 10, Werner and Dindal, 1990; 11, Paoletti et al., 1989; 12, Favretto et al., 1992; 13, Stinner et al., 1993; 14, Lockeretz, 1987; 15, Karg, 1989; 16, Exner et al., 1990; 17, House and Rosario-Alzgaray, 1989.

Conserving biodiversity in managed crop and pasture systems provides many opportunities as well as many challenges to scientists and planners. The objectives of sustainable agriculture and biodiversity conservation are economically compatible and should be actively pursued (Altieri et al. 1987; Reid and Miller, 1989; Pimentel et al., 1991; Paoletti et al., 1991; Huston, 1994).

6. Using species as tools for measuring farming systems sustainability

Plant species abundance have been suggested as a tool to monitor landscape biodiversity (Huston, 1994). However, in rural landscapes animal species, especially invertebrates, are better than plants species in abundance and flexibility, and seem to be a more available tool to use (Paoletti et al. 1991; Holloway and Stork, 1991; Paoletti and Pimentel, 1992; Heliovaara and Vaisanen, 1993; Pimentel, 1994).

We wish to show a few examples of how invertebrates can be used to detect environmental stresses of some agriculture practices, giving useful information on agroecosystem sustainability.

In the Toscana region we compared some vineyards with different cultivation (tillage vs. no tillage) and mulching (bare soil vs. subterranean clover living mulch). Fig. 2 shows the distribution of mesofauna (mites, insect larvae, springtails, etc.), collected with the Tullgren modified extractor. We found a higher number of mesofauna invertebrates in the mulched vineyards, which are the less disturbed fields (Favretto et al., 1992).

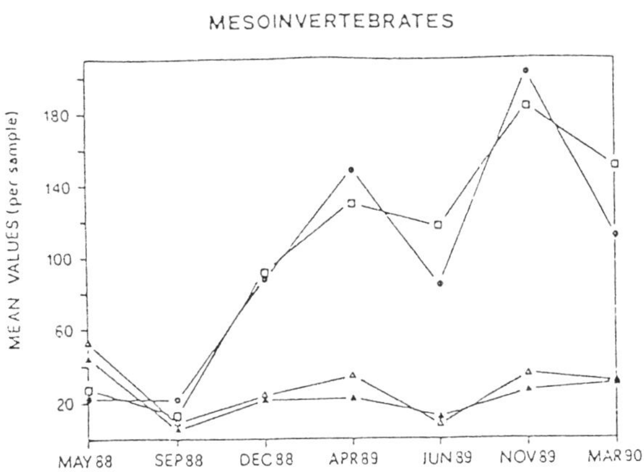

Fig. 2 Mean values of Mesostigmata collected by modified Tullgren system in different vineyards in central Italy: □, mulched plots with fertiliser; O, mulched plots without fertiliser;Δbare plots with fertiliser; ▲ , bare plots without fertiliser.

Invertebrates react to general impacts of pesticides, fertilisers and tillage in two different kinds of orchards (organic and conventional). In particular detritivore (earthworms, millipedes, woodlices, some insect larvae) density are depressed in conventionally grown apple orchards if compared with organic ones, Fig 3.

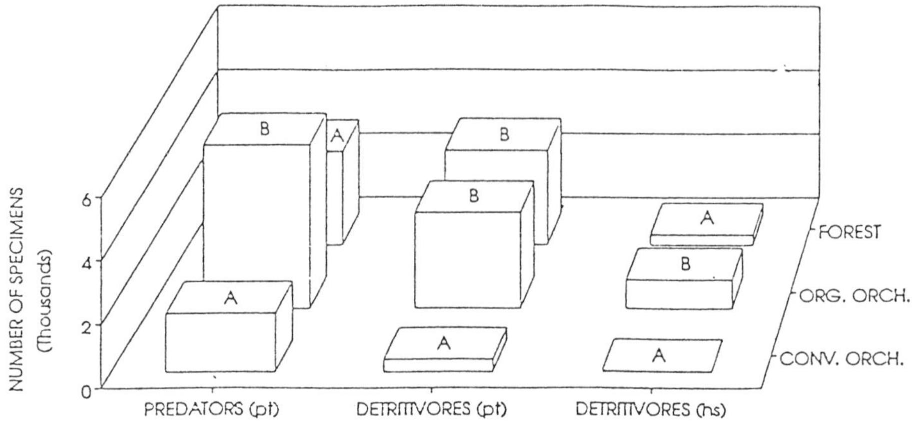

Fig. 3. Total specimens of detritivores and predators collected monthly in ten pitfall traps, from April to November 1987, and total specimens of detritivores collected in hand sorted cores (30 cm x 30 cm, five cores x four collections). Predators collected in pitfall traps are Chilopoda, Araneae, Opiliones and Carabidae. Detritivores collected in pitfall traps are Diplopoda and Isopoda: detritivores colected by hand sorted cores are Lumbricidae, Diplopoda and Isopoda (from Paoletti et al. 1995b).

During 1990-1991 we compared six peach orchards with different agricultural practices (biological, integrated and conventional). Invertebrates were sampled using pitfall traps, the Tullgren modified extractor and sweeping nets. Many groups of beneficials invertebrates, such as Carabidae, Araneae, Diplopoda, Isopoda, etc., had been classified to species. Classified organisms represented only a small part of total biodiversity, but can give some useful information. Total species number was higher in biological farm compared with conventional ones; integrated farms presented intermediate figures between biological and conventional ones (Fig. 4). Integrated farms presented a reduction of 15 % in species number, while conventional ones had 35 % lower species than biological farms. Some group in particular showed an interesting response to agricultural input: for example spiders number of species was 49-50 in biological, 36-44 in integrated and 29-30 in conventional farms, with a reduction in the latter of 39-42 %. The number of Braconidae species was 8-10 in orchards with spontaneous vegetation, while it was only 1-2 in tilled orchards. The terrestrial crustacean isopods had only a few more species in biological farms than in conventional one (5 vs. 3-4) but their total activity density reflect agricultural input, decreasing from biological to integrated and conventional farms. Some species were found only in biological and sporadically in integrated farms, but they were absent in conventional ones; such in the case of the spider Pachygnatha degeeri and the ground beetles Brachynus sclopeta, Campalita auropunctata and Platynus dorsalis (Fig. 5).

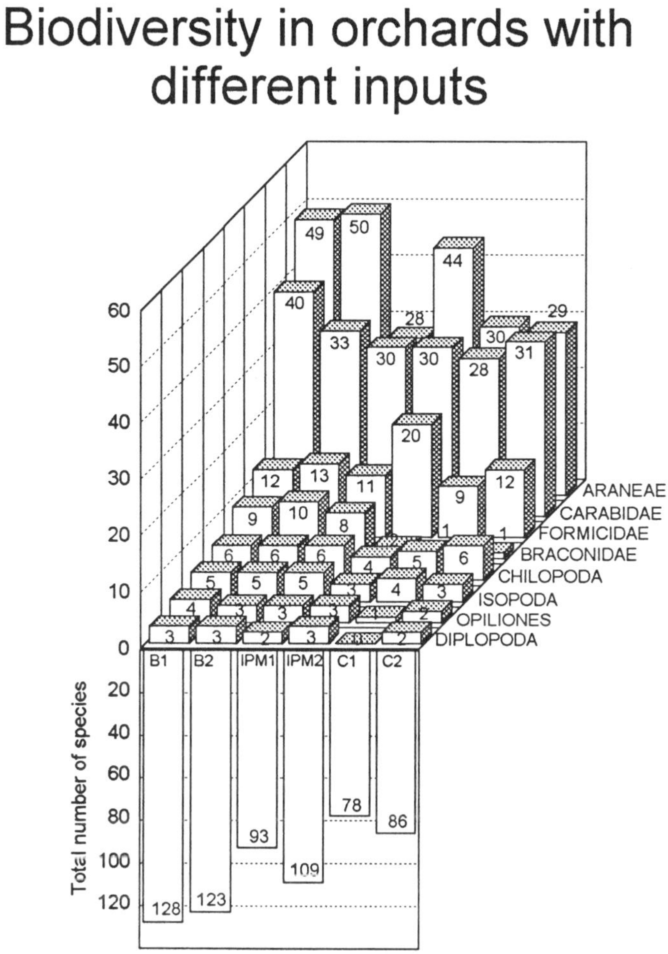

Figure 4. Species number in six orchards with different input: biological (B1 and B2); integrated pest management (IPM1 with soil vegetation and IPM 2 with soil tillage) and conventional (C1 and C2).

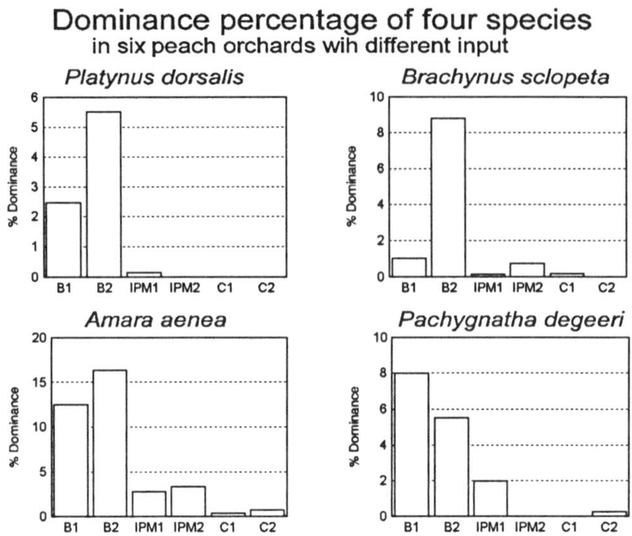

Figure 5. Dominance percentage of some Carabidae (Platynus dorsalis, Brachynus sclopeta and Amara aenea) and Araneae (Pachignata degeeri) species in six peach orchards with different management (see Legend Fig. 4). Number are the means of two sampling years (1990 and 1991). Invertebrates were collected using pitfall traps, six for each orchard.

A common practice in orchard is the use of copper sulphate as fungicides. This compound is usually considered to have low toxicity, even if sometimes, at very high concentration, it can produce some damage to plants (Fregoni and Bavaresco, 1984). We studied the response of earthworm to copper concentration in the soil. In a biological farm we compared the earthworm distribution in three vineyards of different ages and in two control fields. Total earthworm density was reduce by the high copper concentration present in old vineyards and in particular some species, such as Aporrectodea caliginosa, Allolobophora chlorotica, almost disappeared from vineyards (Fig. 6)(Paoletti et al., 1995).

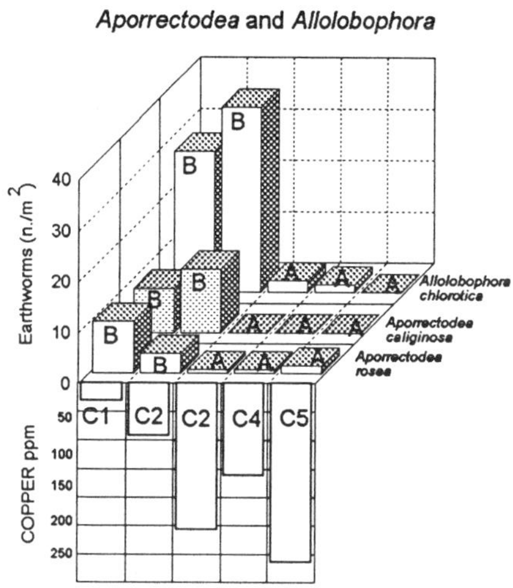

Figure 6. Abundance distribution of some earthworm species in field with different copper input. the lower part of the graphics illustrates the copper concentration in each field. Legend: C1, control field; C2, vineyard from 1930 to 1960, alfalfa field in rotation during the sampling data; C3, vineyard from the beginning of the century until 1930 and from sixties until todays; C4, vineyard 30 years old; C5, vineyards 60 years old (from Paoletti et al., 1995c).

We also compared the earthworm community in 72 orchards belonging to 4 groups characterised by different chemical input: apple orchards (high input); vineyards (medium input but high copper supplied); peach orchards (medium input) and kiwi-fruit orchards (low chemical input). In each group, orchards were characterised by different soil cultivation (tillage vs. no tillage). Earthworm communities were severely reduced by copper concentration, more than any other chemical compounds studied. In particular 'endoges' group (earthworms living and feeding in deep soil layer) disappeared when copper concentration in soil was higher than 170-180 ppm and their distribution in orchards was explained by copper concentration in soil for 46 %. On the other hand soil tillage severely affected the presence of epigeic earthworms (species living and feeding in the upper soil layer). In this case, earthworms can give useful information on farm management, allowing to assess separately chemical and mechanical disturbance (Fig. 7).

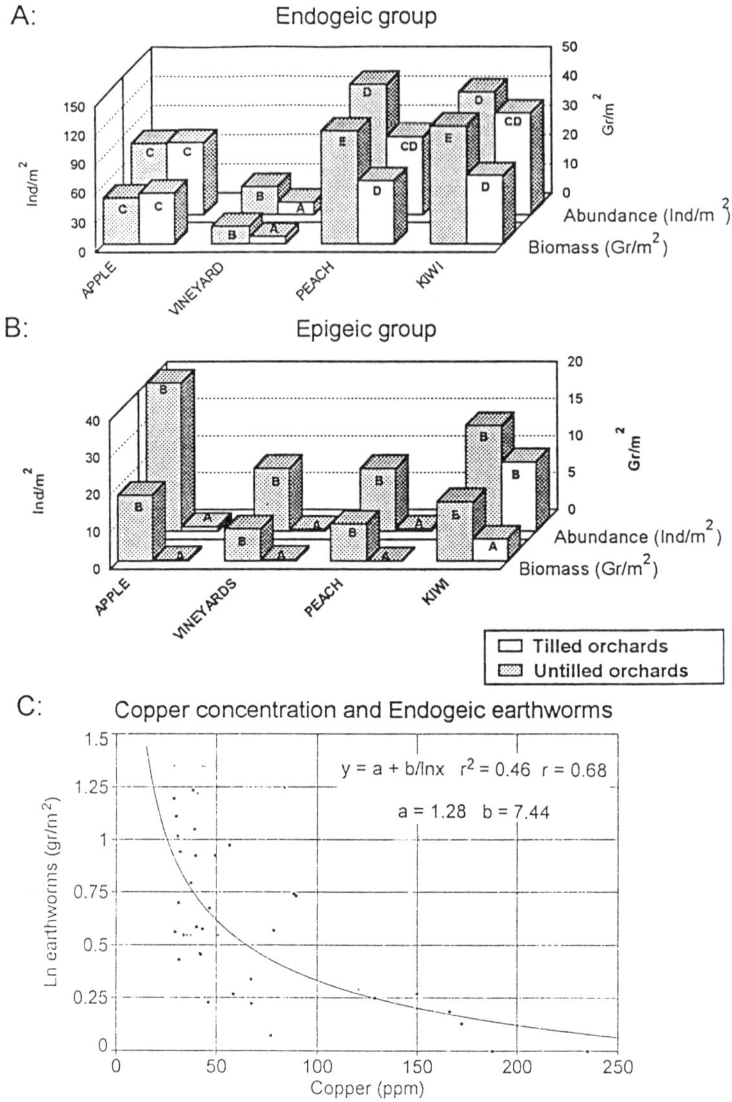

Figure 7. (A) Distribution of endogeic earthworms in 64 orchards with different chemical input: apple orchards with high input; vineyards with medium chemical input but high copper supply; peach orchards with medium chemical input but low copper supply; kiwi orchards with low chemical input. (B) Distribution of epigeic earthworms in orchards with chemical input; see below. (C) Regression between endogeic earthworms and copper content in soil. Each point represents one orchard and the values are the mean of 5 samplings for both earthworms and copper concentration.

TABLE 3: some recent efforts developed in order to analyse and evaluate agroecosystems and rural landscapes are summarises, using invertebrate biomass or biodiversity, mostly relating to soil invertebrates (F\from Paoletti 1995.

Animal	Landscape	Location	Colletion system	Ref.	Results
Earthworms	Woodlands-rural	NE Italy	Handsorting	Paoletti, 1988	+o
Earthworms, carabids, diplopods, isopods, araneids	Vineyards/corn/ alfalfa/meadows	NE Italy	Handsorting, cores and pitfall traps	Paoletti et al., 1988	+o
Soil organisms	Apple orchards	N. Italy	Cores, pitfall	Paoletti et al., 1993	+o
Soil and aboverground invertebrates	Peach orchards	N. Italt (Forlì)	Pitfall traps, hand sorting, cores	Paoletti et al, 1993	+o
Soil meso-invertebrates	Vineyards inputs	Central Italy	Soil cores	Favretto et al, 1992	o
Protozoa, earthworms	Agroecosystems, input	Austria	Soil cores	Foissner, 1992	o
Earthworms	Different input agroecosystems	USA	Soil cores	Werner and Dindal, 1990	o
Invertebrates	Different input agroecosystems	Switzerland	Soil cores, pitfall traps	Matthey et al., 1990	o
Carabid	Different input agroecosystems	Austria	Pitfall traps	Kromp and Steinberg, 1992	o
Vertebrates	Disturbed/undisturbed forests	North America		Hansen et al., 1991	+o
Flying invertebrates	Different landscapes	Poland, Italy Russia		Karg, 1989 Ryszkowski et al, 1991	+o
Spiders, Hymenoptera	Wetlands and others	Switzerland	Pitfall, ecc.	Duelli et al., 1991	+
Carabids and spiders	Hedgerows presence/absence in agroecosystems	NE Italy	Pitfall traps	Nazzi et al., 1989	*X
Carabids and spiders	Hedgerows presence/absence in agroecosystems	NE Italy	Pitfall traps	Schirato, 1991	*X
Carabids and spiders	Hedgerows presence/absence in agroecosystems forest	NE Italy France	Pitfall traps Pitfall traps	Targa, 1990 Burel and Baudry, 1990	*x

+, more species and less input; *, more species near hedgerows; o, higher biomass and less input; X, more biomass near hedgerows.

Conclusions

For several years, invertebrate communities have been used as bioindicators to detect environmental health in aquatic ecosystems (Woodiwiss, 1978; Lebrun, 1981; Blandin and Lamotte, 1985; Blandin, 1986). In recent years there has been an increasing need for tools that can give information on soil pollution and environmental stress in terrestrial ecosystems, and invertebrate biodiversity can be very useful in assessing the situation (Paoletti et al., 1991; Pimentel, 1994).

A limited number of extensive research plans are not sufficient to address general views about sustainability and number of species carrying capacity. Hoewer if we extrapolate our data for peach orchards we can suppose a reduction of total biodiversity of almost 35 % in conventional versus biological farms. Some invertebrate groups or species can give more information on environmental management; for example, spiders, ground beetles, isopods, etc. Earthworms in particular seem very useful bioindicators because they are relatively large animals, usually easy to collect (a simple spade may be sufficient) and are very sensitive to soil condition. Moreover if earthworm are considered at species or morphoecological levels (Bouchè, 1977; Lee, 1985) (this last one is usually easy to recognize) they can give information not only on total environmental stress but also on single agricultural input (such as copper pollution and soil tillage).

Today much research has evaluated the effect of individual chemical compound on soil invertebrates, but much of this research has been developed in laboratory conditions. For example Edwards and Bohlen (1992), in their review of the effect of chemical compound on earthworms recorded 609 laboratory studies but only 300 field studies.

More research is needed to better understand the effect of individual agricultural practices on the invertebrate community, both as total biodiversity scale and as single component, in order to obtain a framework of information that can be used not only by scientists but also by farmers, agricultural technicians and politicins in their environmental management programs to improve sustainability.

References

Altieri, M.A., Anderson, M.K. and Merrick, L.C. (1987) Peasant agriculture and the conservation of crop and wild plant resources, *Conserva. Biol.*, **1**, 49-58.

Blandin, P. (1986) Bioindicateurs et diagnostic des systemes ecologiques. *Soc. Ecol.*, **17**, 1-307.

Blandin, P. and Lamotte, R. (1985) Ecologie des systemes et amenagement: foundaments theoriques et principes methodologiques, in R. Lamotte (ed.), *Fondaments rationnels de l'amenagement d'un territoire*. Masson, Paris, pp. 139-162.

Bouchè, M.B. (1977) Strategies lombriciennes, in U. Lohm and T. Persson (eds.), *Soil organisms as components of ecosystems*, Biological Bolletin, Stockholm, 25, 122-132.

Buszko, J. (1991) Motyle (Lepidoptera) rezerwatu Las Piwnicki, *Parki nar. Rez. przyr.*, Bielowieza, **10**, 5-60.

Clawson, D.L. (1985) Harvest security and intraspecific diversity in traditional tropical agriculture, *Economic Botany*, **39**, 56:67.

DeFolliart, G. (1989) The Human Use of Insects as Food and as Animal Feed, *Bull. Entom. Soc. of America* **35**, 22-35.

DeFolliart, G. (1990) Insects as food in Indigenous Populations, in D.A. Posey, W.L. Overnal, C.R. Clement, M.J. Plotkin, E. Elisabetsky, C. Novaes da Mota and J.F. Pessos de Barros (eds.), *Ethnobiology: Implications and Applications*. Museum E. Goeldi, Manaus, pp. 145-150

Dover, N. and Talbot, L. (1987) To Feed the Earth: Agroecology for Sustainable Development. World Resources Institute, Washington, DC.

Edwards, C.A. and Bohlen, P.J. (1992) The effects of toxic chemicals on earthworms, *Reviews of Envrironmental Contamination and Toxicology*, **125**, 23-99

Ehrlich, P.R. and Wilson, E.O. (1991) Biodiversity Studies: Science and Policy. *Science*, **253**, 758-762.

Erwin, T.L. (1982) Tropical forests: Their richness in Coleoptera and other Arthropod species. *Coleopt. Bull.*, **36**, 74-75.

Favretto, M.R., Paoletti, M.G., Caporali, F., Nannipieri, P., Onnis, A. and Tomei P.E. (1992) Invertebrates and nutrients in a Mediterranean vineyard mulched with subterranean clover (Trifolium subterraneum L.), *Biol Fertil Soils*, **14**, 151-158.

Fregoni, M. and Bavaresco L. (1989) Il rame nel terreno e nella nutrizione della vite, Vignevini, **5**, 37-49.

Hansen, A.J., Spies, T.A., Swanson, F.J. and Ohmann J.L. (1991) Conserving Biodiversity in Managed Forests. *BioScience*, **41**, 382-392.

Hawkes, J.G. (1990) The potato Evolution, Biodiversity and Genetic Resources. Smithsonian Institution Press, Washington DC.

Heliovaara, K. and Vaisanen R. (1993) Insects and Pollution. CRC Press, Boca Raton Fl.

Holloway, J.D. and Stork, N.E. (1991) The Dimensions of Biodiversity: The use of Invertebrates as Indicators of Human Impact. In: D.L. Hawksworth (ed.), *The Biodiversity of Microorganisms and Invertebrates: Its Role in Sustainable Agriculture*. CAB International, London, pp. 37-61.

Huston, M.A. (1994) Biological Diversity: The Coexistence of Species on Changing Landscapes. Cambridge Univ. Press, Princeton.

Lebrun, P. (1981) L'usage des bioindicateurs dans la diagnostic sur la qualitè du milieu de vie, in Ecologie appliquèe: indicateurs biologiques et techniques d'ètudes, Journèes d'ètudes, Grenoble, 13-14 novembere 1980. Association Française des Ingénieurs Ecologues, Mainvilliers, pp. 175-202.

Lee, k.E. (1985) Earthworms: their ecoloy and relationships with soils and land use. Academic Press.

May, R. (1992) How many species inhabit the earth? *Scientific American*, **267**, 42-48.

Michon, G. (1983) Village-forest-gardens in West Java. In: A.Huxley (ed.) *Plant Research and agroforestry: proceedings of a consultative meeting held in Nairbi*, 8 to 15 April, 1981. pp. 13-24.

Ogle, B.M. and Grivetti, L.E. (1985) Legacy of the Chameleon: edible wild plants in the kingdom of Swaziland, southern Africa. Cultural, ecological, nutritional study. Part IV. Nutritional analysis and conclusions. *Ecol. Food Nutr. N.Y.*, **17**, 41-64.

Paoletti, M.G. (1995) Biodiversity, traditional landscapes and agroecoystem management, *Landscape Ecology and Urban Planning*, 31, 117-128.

Paoletti, M.G., Dreon A.L. and Lorenzoni G.G. (1995a) *Pistic*, traditional food from western Friuli, N.E. Italy, *Economic Botany*, **49**, 26-30.

Paoletti, M.G., Favretto, M.R., Stinner, B.R., Purrington, F.F. and J.E. Bater (1991) Invertebrates as bioindicators of soil use. *Agric., Ecosys. Environ.*, **34**, 341-362.

Paoletti, M.G., Favretto, M.R., Ragusa, S. and Strassen, R. (1989) Animal and plant interactions in the agroecosystems. The case of woodland remains in north-eastern Italy, in: F. Golley and P.M. Golley (eds), *Environmental Consequences of Agricultural Policy and Practice*. Ecology International, Bull. **17**, 79-91

Paoletti, M.G. and Pimentel, D. (eds.) (1992) Biotic Diversity in Agroecosystems, Elsevier, Amsterdam.

Paoletti, M.G., Pimentel, D., Stinner, B.R. and Stinner D. (1992) Agroecosystem biodiversity: matching production and conservation biology. *Agriculture, Ecosystems and Environment*, **40**, 3-23.

Paoletti M.G., Schweigl, U. and Favretto M.R. (1995b) Soil macroinvertebrates, heavy metals and organochlorines in low and high input apple orchards and a coppiced woodland, *Pedobiologia*, **39**, 20-33.

Paoletti, M.G., Sommaggio, D., Petruzzelli, G., Pezzarossa, B. and Barbafieri, M. (1995c, in press) Soil invertebrates as monitoring tools for agricultural sustainability, *Polskie Pismo Entomologiczne*, **62**.

Pimentel D. (1994) Insect population responses to environmental stress and pollutants, *Envrironmetal Research*, **2**, 1-15.

Pimentel, D., McLaughlin, L., Zepp, A., Lakitan, B., Kraus, T., Kleinmann, P., Vancini, F., Roach, W.J., Graap, E., Keeto, W.S. and Seling, G. (1991) Environmental and economic impact of reducing US agricultural pesticide use, in D. Pimentel (ed.), *CRC Handbook of Pest Management in Agriculture*. CRC Press, Boca Raton, Fl., pp. 679-718.

Pimentel, D., Stachow, U., Takacs, D.A., Brubaker, H.W., Dumas, A.R., Meaney, J.J., O'Neil, J., Onsi, D.E. and D.B: Corzilius (1992) Conserving Biological Diversity in Agricultural/Forestry Systems, *BioScience*, **42**, 354-362.

Reid, W.V. and Miller, K.R. (1989) Keeping Options Alive. The scientific basis for conserving biodiversity. W.R.I. Washington D.C.

Stork, N.E. (1988) Insect diversity: facts, fiction and speculation. *Biol. J. Linn, Soc.*, **35**, 321-337.

Wheeler, Q. (1990) Insect Diversity and Cladistic Constraints. *Ann. Entomol. Soc. Am.*, **83**, 1031-1047.

Wilson, E.O. (1988). Biodiversity. National Academic Press, Washington D.C.

Woodwiss, F.S. (1978) Biological water assessment methods. Severn Trent River Authoritie, Commission of the European Communities, Bruxelles.

AN INTEGRATED BIOINDICATION SYSTEM APPLIED TO SOIL POLLUTION ASSESSMENTS: FROM EARTHWORMS TO ECOSYSTEMS

M.B. BOUCHÉ
Laboratoire de zooécologie du sol (INRA/CNRS)
BP 5051, F-34033 Montpellier Cedex I
France

Abstract

The proposed bioindicator system is based on an integration of the information gathered on earthworms in polluted sites. This integration deals with pollutant effects on survival (presence-absence), tissue concentrations (body burden of heavy metals, PCBs, PAHs, genotoxics, *etc.*) and on effects on the earthworm community. The integration deals also with pollutant assessment on the ecosystem resulting from community changes. These ecosystem assessments concern physical (erosion, overflow, drainage, soil stability), chemical (carbon, nitrogen cycles, pollutant fate, *etc.*) and biological properties (food chains, biocontaminations, *etc.*). The bioindicator system integrates these elements by modelling and a description of the state-of-the-art in an Explained Knowledge Dispenser. With this approach, up-to-date knowledge could be accessible and shared by integrology between the various research teams and decision-makers. It will allow to make predictions *versus* time, in a falsifiable manner allowing permanent improvements.

1. Introduction

Framework. A pollutant is a substance, simple or complex, having ecotoxicological effects, *i.e.* a toxicity on organisms living in ecosystems. Earthworms constitute the main animal biomass in our terrestrial ecosystems (110 g live weight m^{-2} in France; n = 62). They are present nearly everywhere and, depending on their ecological categories, dwell in almost all soil compartments with living components, from the epiphytic soils in tropical rain forests to the deep soil layers (2 m or more). Therefore earthworms are good candidates for assessments.

N. M. van Straalen and D. A. Krivolutsky (eds.), Bioindicator Systems for Soil Pollution, 141–153.

To assess a pollutant, we must observe in soils its effects on ecosystems and human welfare. This is not enough. To assess pollutant risks, *i.e.* biological and ecosystemic disorders, the information gathered must be ordered and accessible to other specialists and decision-makers. To order knowledge two linked backgrounds are needed: modelling and integrology.

Modelling. This has been improved since the IBP [1], but drawbacks have led me to formalize model parameters: 1) The model components must be measurable in true systems, *i.e.* (agro)ecosystems, 2) Results must be falsifiable in ecosystems, 3) The model must be transposable *(i.e.* usable in a lot of ecosystems with a minimum of local variables), and 4) The model must be a synthetic tool, *i.e.* designed to be, simultaneously, a submodel of ecosystems and to integrate submodels fitted to some of its subcomponents.

Such a model was designed in 1975 and named REAL (Rôle Écologique et Agronomique des Lombriciens = Agronomical and Ecological Earthworm Role) [2, 3, 4], see Fig. 1. This conceptual model currently has numerous submodels (numerical, qualitative and fuzzy models). The effects of pollutants are registered at the individual level: stage, state, species, and if needed, body and *endentere* chemical composition (*endentere* = digestive tract content). In a given site these data are aggregated to describe effects on population and ecological categories. The ecological category concept [5, 6] allows to transpose REAL from one site to other sites, with changes in soil, species, climatic and human management conditions. Only local population studies and, if possible, soil temperature and moisture are needed to feed the model. REAL allows for the description of changes in ecosystem function in terms of physical variables (water infiltration rate, rate of soil stable crumb formation) [7, 8, 9], chemical variables, especially nitrogen dynamics [10, 11, 12] and heavy metal content of earthworms as food for other animals [13, 14, 15]. Presently, REAL does not allow for the prediction of rates of population restoration and food chain transfer of contaminants, because of the lack of demographic modelling.

Integrology. The second background, integrology, [16] was founded when the concept of **relation,** which has been intensively used in Relational Data Bases since 1984, met clear ecological concepts as DICs, prelevat, referender, described elsewhere [4] (Ecological means here globally biophysicochemical concepts, not biological or chemical or physical ones). This was intensively improved and used for an integrated management of biophysicochemical data (with no discipline borders) especially in the Relational Data Base BASECOL of ECORDRE, where billions of data (DICs) belonging to more than one thousand variables are integrated and automatically accessible [4, 17].

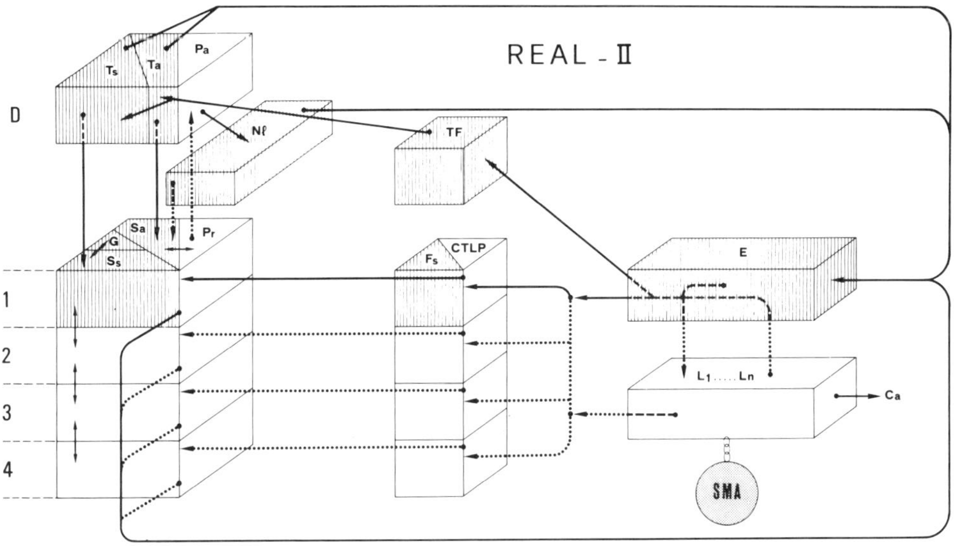

Figure 1. The REAL Model describes the roles of earthworms in ecosystems, which depend of the population levels (L_1, L_2, ..., L_n), affected or not by pollutants. Earthworms act on ecosystems by ingestion of: 1) mineral (Ss) or organic (Sa) soil, creating burrows (G), 2) their old faeces (Ts and Ta), and 3) the litter (Nl). The digestion of the endentere (E = digestive tract content) produces assimilation to L or faeces production (Fs) with a quick transfer to plant (Pr, Pa), through a transitory compartment (CTLP). This function varies with soil layers (1, 2, 3, 4, *etc.*) and the level of earthworm activity described by the submodel of activity (SMA). The pollutant body burden is transfered to predators (Ca). From [4].

More recently, the concept of relation became practicable to interpret DICs at all levels of cognition (from very tiny and sophisticated focused studies to interpolations, from highly validated results to risky falsifiable hypotheses). The description of knowledge elements integration is in progress in an Explained Knowledge Dispenser, ECOGNIT of ECORDRE, which includes a subdivision called ROLUMBRIC. We are describing, in ROLUMBRIC, REAL and its relation with assessments of pollutants. The EKD ROLUMBRIC [18] will give in full extended access to interpreted knowledge, then to the process of interpretation. It opens to an on-line knowledge service with users (*i.e.* other specialists or decision makers). Users will be able to use their local data, to ask for explanations, and make criticisms on ROLUMBRIC. ROLUMBRIC, of ECORDRE, is a tool which will be accessible on INTERNET soon.

2. The proposed new bioindicator system

2.1. PRINCIPLES

The proposed new bioindicator system is based on two principles:

1) *It is an open svstem of information*, convivial, interactive and giving access to knowledge in a way adapted to users. Giving elements of knowledge, these elements are selected by each user for its needs and are not served pre-structured, following another need or mood.

2) *It is not pre-oriented* by a favored hypothesis: the bioindicator is not to be used to indicate definitively something (example: the presence of this species means the soil is clean), but it is just used for the information it bears including temporary proposed interpretations.

Interpretations are temporary, or extemporaneous, because they are always submitted to criticism and reinterpreted for improvements. A bioindicator indicates one or several properties as hypotheses not as absolute truths (there is no room for truth in science); these hypotheses, which depend widely on local conditions, must be used carefully, *i.e.* knowing elements of knowledge used to build them and assessing the level of uncertainty they bear. For this reason a bioindicator here is in the first place a bio-informant.

The bioinformation gathered on earthworms is managed following integrological principles, including modelling, as indicated as fundamental backgrounds above.

2.2. INFORMATION DERIVED FROM EARTHWORM ECOLOGY

The information derived from earthworm ecology is of three types.

1) The direct information from them in ecosystems (= their state variables) allows **biomonitoring**, including in experimental plots.

2) The integrated information on their relations with other (agro)ecosystem components allows **environmental assessments**.

3) The modelling versus time of population changes (*i.e.* demographic and health state changes) and their related ecosystem consequences, allow **predictions**.

2.3. BIOMONITORING

Collection of knowledge. The quality of the information gathered from the field, using appropriate sampling methods and techniques depends on the objective of the study. The precise methodology has been described elsewhere [4, 19, 20].

The sampling of earthworms could be organized to get **information on individuals** and on other appropriate variables (soil, plants, *etc.*). This is made efficiently by punctual methods (linking of the various prelevats giving informations by a common unique spatio-temporal point, see [19]) and by formless prelevats (a prelevat is a formalized sample unit, see [4, 19]). This implies, for example, sampling by means of a spade in a sample unit designed in agreement with the features of the environment and not fixed by the standardized dimensions of the prelevat.

The sampling of earthworms could be designed to get, in addition, information on **community level and structure**. To obtain this information, it is necessary to use: 1) the stational method with euclidian (or cartesian) prelevats, *i.e.* unit samples having tridimensional (or bidimensional) size dimensions. These dimensions allow further interpolations at station (field) level, and 2) a quantitative extraction technique of earthworms from soils. The different techniques, either physical, ethological or physical, are tedious and time consuming, and their efficiency depends on local factors [20, 21]. This stational method associated with quantitative extractions works well with plot studies.

The punctual method associated with formless prelevats is far more flexible, efficient and cheap than the stational method with euclidian prelevats and should be used if knowledge about community structure and level is not absolutely needed. "Intermediate" practices *(e.g.* punctual method with euclidian sampling) must be discarded: they increase cost with no gain of information.

Analysis of prelevats must be made on each individual for general features (stade, state, species, weight, shape, *etc.*). Body chemical and molecular biochemistry analyses suggest it is important to clear the digestive tract. Various techniques (dissection, cellulose or levilite feedings are discussed elsewhere: [22, 23]).

All analytical data must be registered and managed, if possible, in an accessible Relational Data Base linking each DIC (date, initial and controlled) to other DICs to connect them to the five referenders: time, space, composition *(e.g.* characteristics), protocol (of analyses and observations) and observer (who did it, for who, data gathering).

The interpretation of bioindications. The presence of a susceptible taxon can indicate a clean soil if the taxon observed is well known to be susceptible to pollutants [24]. The reverse (absence) is just an assumption of risk: the absence could be because of another factor, such as an inadequate soil pH.

Earthworms eat soil components: minerals, dead organic matter from plants, animals and microorganisms. They digest and assimilate from this "sample". This sample is taken from the "live" part of the soil *i.e.* the top soil layers, the rhizosphere, and the burrows deeper in the soil. The sample does not deal with the majority of deep horizons which are inert (with no life activity). Earthworm have quick turnover of nutrients (10 % per day of their nitrogen is assimilated and excreted), and their body levels of pollutants reflect, most probably, the true ecotoxicological level of the soil (= the true bioavailable fraction of pollutants). A toxicological level is related to life and not to an unassimilable fraction as for soil analysis [15]. Earthworms are biosampling for us the biocirculating fraction of pollutants. This can be determinated for heavy metals [25, 26, 27], PCBs [23, 28, 29], PAHs [28, 30] and, more recently, for genotoxics by subquantification of adducts on earthworm DNA [31]. The level of these biorisks could be mapped or followed *versus* time.

The quantitative impact of a pollutant could be measured by the estimation of community levels and structures subjected to soil contamination either intentionally *(e.g.* field plot) or accidentally, by comparison with a non-contaminated soil. That is the unique ecotoxicological trial we have. Changes in population levels allow for estimation of most ecological consequences.

2.4. ASSESSMENTS OF CONSEQUENCES

The quantitative impact of a pollutant could be quantitatively measured by field or plot comparisons of earthworm population size. These population changes could be related to changes in ecological categories, and the latter changes reported to the earthworm functions estimated by REAL submodels. These changes could be transitory: 1) if the populations are not totally destroyed and an inoculum survives to restore the population, or 2) if the pollutant is ephemeral (degradable, *etc.*). The assessment of soil restoration depends on our ability to predict both pollutant persistence and the capability for a particular population to recover (see section 2.5.: Predictions).

Changes in earthworm communities may be evaluated for three types of consequences.

Soil physics. Especially soil stability and vertical water infiltration are important in this respect. In France, a mean population of 100 g m^{-2} live earthworms produces 30,000 g m^{-2} dry weight of stable crumbs and preserves a mean infiltration rate of 160 mm of water per hour, *i.e.* about one fifth of the annual rainfall [7, 8, 9].

Ecosystem chemistry. Especially the nitrogen cycle is of importance here. Direct soil observations have allowed us to estimate, for a mean population in France, an ingestion rate of 23,000 g N m^{-2} per year, with an excretion rate of 5300 g N m^{-2} which is almost totally assimilated by plants within a few days during the growing season while only 30 % of it is assimilated in September [10, 11, 12]. Effects on the carbon cycle are not so well described [10], but processes have a rate which is 4 to 5 times higher.

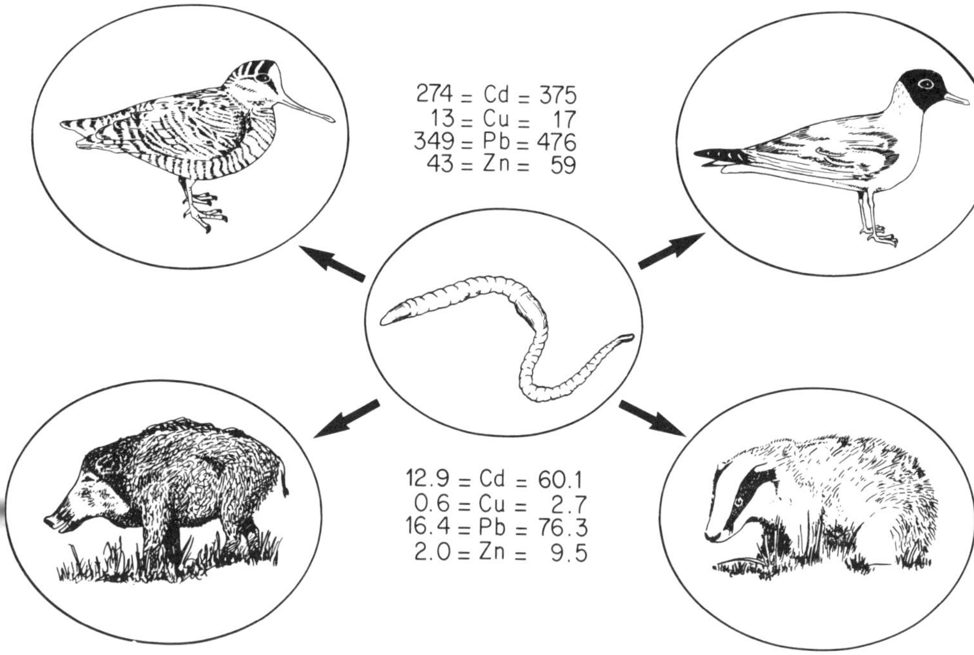

Figure 2. Transfer of heavy metals from earthworms to four important predators: the woodcock, the black-headed gull, the pig and the badger. For each predator, the number indicates the ratio between the estimated daily consumption of metals and the maximum acceptable daily intake for the human population. The data are estimated from 126 samples taken in South-East France [37].

Biological consequences. These are are more poorly documented. A lot of microcosm or pot studies demonstrate the great effect of earthworms on the other two major biological components of ecosystems: plants and microorganisms [32, 33]. Direct evidence was introduced early by the description of fertile ecosystems with mull humus type and earthworms as opposed to poor soil with mor humus type and no earthworms [34] or only epigeic species [35]. This indirect evidence could be hardly quantitatively linked to earthworm level changes.

Earthworms play an important role in food chains of numerous vertebrates by serving as their prey [36]; the consequences of heavy metal body burdens in the diet of four species (two birds and two mammals) have been estimated [38], see Fig. 2. These data show that earthworm predators consume metals at rates exceeding the acceptable daily intake for humans by a factor of 2 to 476, depending on the metal and the predator.

2.5. PREDICTIONS

To predict it is necessary to have a model describing real systems (= ecosystems) *versus* time. The proposed bioindicator system is linked with the conceptual model REAL (Fig. 1). As illustrated in Fig. 3, it can be transposed from field to field and to different earthworm community states, depending on pollutant effects, due to a general method of extrapolation using: 1) local population variables, 2) local soil moisture and temperature data, and 3) the latitude situation to calculate day-length [4]. The various quantitative (mathematical) or qualitative submodels of REAL are linked and described in the EKD ROLUMBRIC (EKD = Explained Knowledge Dispenser, an on-line knowledge service). Presently, two major drawbacks remain: 1) the absence of integration of our knowledge on the estimation of the pollutant persistence, and 2) the absence of demographic models predicting the earthworms' ability to restore their population levels after pollution exposure. Despite the use of various matrix models (Leslie, Lefkovitch, Lexis, Hadjibiros) [38] and the management of the greatest earthworm population study *versus* time [20] we were unable to produce a demographic model falsifiable based on field data. The improvement of field ecophysiological models [39] and fuzzy modelling (see [4], and G.P. Stamou, chapter 6 of this book) will lead to such falsifiable models. At present, we must sample *versus* time to follow the community restoration (T in Fig. 3).

Earthworms are not only used as bioindicators in ecosystems but also as tools in ecotoxicity tests, to provide decision makers with scientific information supporting the homologation of new chemicals before marketing [40], or to manage pollutants, such as industrial waste [41] prior to disposal or land-spreading. The bioindication system could be completed by developing links with these laboratory tests.

3. Example of application

The use of earthworms as bioindicators for the three aims described above (biomonitoring, impact assessment, prediction) is well accepted and these animals are commonly used for the assessment of soil pollutants. Examples of applications, at the research level of implementation, are numerous and are well documented in classical textbooks [32, 33, 42, 43].

Despite the relatively high number of studies this tool is poorly used in pollution assessment because these studies are scattered in the literature and are poorly accessible due to a lack of integrology [16], blocking the sharing and improvement of the new bioindicator system proposed here. In fact, this new system forces us to integrate knowledge *versus* conceptual models, such as REAL, which must be criticised and improved in real time.

The integration, *i.e.* their quick access, of the elements of knowledge we are gathering around the world about the theme of "earthworms as pollutant bioindicators" must be first improved. The gathering of elements of knowledge without their full access is similar to the collection of water with a sieve: while we are collecting the elements we lose the resource.

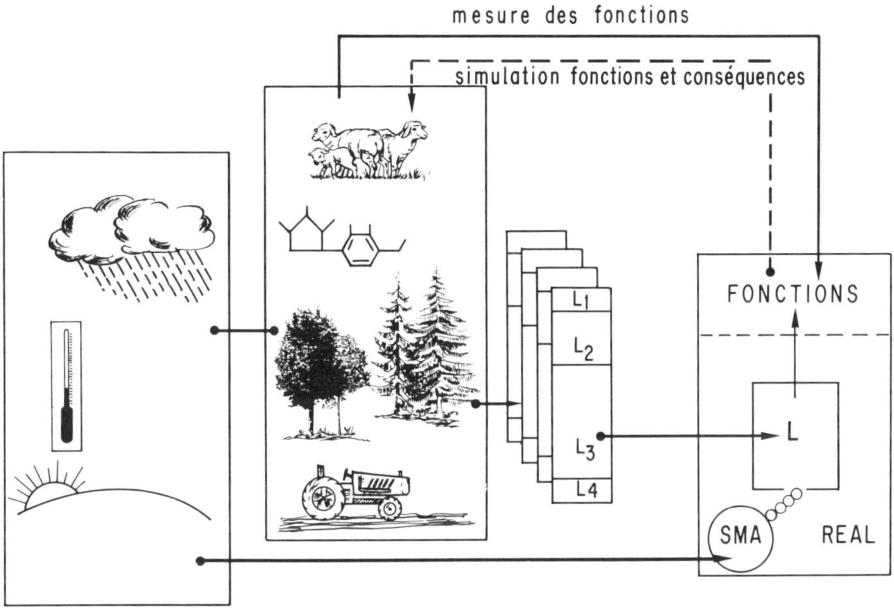

Figure 3. Transposability of REAL in various field conditions (grassland, forest, culture, *etc.*) under various pollutions leading to earthworm community states (populations L_1, L_2, ..., L_n), changing *versus* time (T). At each local condition the various earthworm functions depend on local variables: community state and activity of earthworms. The activity of earthworms depends of the Sub-Model of Activity (SMA) which depends mostly on temperature and soil moisture and on day-length (photoperiod). Most earthworm functions were estimated in fields and are simulated for fields. From [4].

4. Discussion

We are leaving the end of a historical period in science: the need to choose between increasing specialization and efficiency on the one hand, and fuzzy generalisations on the other. There is a conflict between the need to know precisely the effects of a pollutant under defined local conditions and the absence of a multivariate assessment into the variable global use of the same contaminant.

Due to the high diversity of pollutants, the great variety of their dispersal mechanisms and fate in the environment, the complexity and variability of soil components, including organisms, the use of organisms as bioindicators seems impossible or highly risky. This complexity must be first accepted and managed with its consequences: a bioindication could be hardly univocal and a bioindicator is mostly a bio-informant on the ecosystem states. This complexity must also be managed in grouping together all of the bio-information we collect about pollutants and ecosystems. This grouping must order the various elements of knowledge we have, and this ordination must allow the retrieval of those elements that are needed to solve each specific problem. Rather than to continue to make specialized studies independently and to exchange only some tiny parts of them as "results" we must absolutely group all our knowledge and share it. This is now possible.

The organization of knowledge could improve pollutant assessments and allow us to recognize the most vacant areas of knowledge, on which, therefore, we must focus our research efforts. Today, among soil animals, earthworms allow us a rather complete set of assessment approaches, with some drawbacks and a general need of validation and elaboration. This is only possible if we focus our research together, and if we open the access of the knowledge to nearly everybody.

5. References

1. Gounot, J. and Bouché, M.B. (1974) Modélisation de l'écosysteme prairial: objectifs et méthodes, *Bull. écol.* **5**, 309-338.
2. Bouché, M.B. and Kretzchmar, A. (1977) R.E.A.L.: un modèle du role écologique et agronomique des lombriciens, *Ecol. bull. (Stockholm)* **25**, 402-408.
3. Bouché, M.B. (1980) Objectifs, compartimentation et faisabilité du modèle REAL (Rôle Écologique et Agronomique des Lombriciens), *Pedobiologia* **20**, 197-211.
4. Bouché, M.B. (1990) *Écologie opérationnelle assistée par ordinateur*, Masson, Paris.
5. Bouché, M.B. (1971) Relations entre les structures spatiales et fonctionnelles des écosystèmes illustrées par le rôle pédobiologique des vers de terre, in H. Pesson (ed.), *La vie dans les sols*, Gauthier-Villars, Paris, pp. 187-209.
6. Bouché, M.B. (1972) *Lombriciens de France. Écologie et Systématique*, INRA, Versailles.

7. Al Addan, F. (1990) Biophysique du sol. Étude quantitative de la régulation, par le travail lombricien, des propriétés structurales en milieux méditerranéens, *Doc. pedozool.* **2**, 1-291 (+ Annexes).
8. Al Addan, F., Aliaga, R., and Bouché, M.B. (1991) Relations entre peuplements lombriciens et propriétés des sols méditerranéens, in G.K. Veeresh, D. Rajagopal, and C. A. Viraktamah (eds.), *Advances in Management and Conservation of Soil Fauna*, Oxford & IBP Publ., New Delhi, pp. 525-537.
9. Bouché, M.B. and Al Addan, F. (in press) Earthworms, water infiltration and soil stability: some new assessments, in *Proc. 5th Intern. Symp. Earthworm Ecology*, Columbus, Ohio, USA.
10. Ferrière, G. (1986) *Mouvements naturels des éléments dans une prairie: quantification des échanges d'azote entre lombriciens, sol et plantes*, Thèse doct. d'État ès-Sciences, Univ. Lyon I, 148 pp. (+ Annexes).
11. Hameed, R. (1989) *Transformation et translocation de l'azote sous influence lombricienne et microbienne: modèles et validation écologiques*, These doct. 3ème cycle, USTL, Montpellier, 156 pp.
12. Al Addan, F., Bouché, M.B., Cortez, J., Heidet, J.-C., Ferrière, G., Mazaud, D., and Samih, M. (in press) The earthworm role in the actual nitrogen cycle, in *Proc. 5th Intern. Symp. Earthworm Ecology*, Columbus, Ohio, USA.
13. Abdul Rida, A.M.M. (1992) Biosurveillance de la contamination du sol: apport de l'étude des lombriciens à l'évolution des risques liés aux elements traces, *Doc. pédozool.* **1** (4), 1-233.
14. Granval, Ph. (1988) *Approche écologique de la gestion de l'espace rural: des besoins de la bécasse à la qualité des milieux*, These doct. 3eme cycle, spéc. Écologie, Univ. Rennes I, 186 pp. (+ Annexes).
15. Abdul Rida, A.M.M. and Bouché, M.B. (1994) A method to assess chemical biorisks in terrestrial ecosystems, in M.H. Donker, H. Eijsackers, and F. Heimbach (eds.), *Ecotoxicology of Soil Organisms*, Lewis Publ., Boca Raton, Florida, pp. 383-394.
16. Bouché, M.B. (in press) *Intégrologie; l'acces aux connaissances en écologie, industrie, agronomie, informatique, environnement, etc.*
17. Soto, P. and Bouché, M.B. (1993) ECORDRE: Bases de données relationnelles pour toutes données écologiques y compris les activités humaines, in *Les systèmes d'information environnementale*, ICALPE Le Bourget du Lac (France), pp. 69-93.

18. Bouché, M.B. (1995) Modern ecology methods and means allowing broadcasting and application of the role of soil fauna knowledge, *Acta Zool. Fennica* **196**, 206-211.
19. Bouché, M.B. (1975) Discussion d'écologie. I. Introduction. II. Obtention de données écologiques et uniformisation spatiale, *Bull. écol.* **6**, 23-32.
20. Bouché, M.B. and Gardner, R.H. (1984) Earthworm functions (Fonctions des lombriciens). VII. Population estimation techniques, *Rev. Écol. Biol. Sol* **21**, 37-63.
21. Bouché, M.B. and Aliaga, R. (1986) Contre une dégradation physique et chimique des sols et pour leur optimisation économique, l'échantillonnage des lombriciens: une urgente nécessité, *La Defense des Végétaux* **242**, 30-36.
22. Chaudonneret, M.B. (1977) *Quelques données sur la composition biochimique des verres de terre, aliment éventuel du bétail et de l'homme*, Rapport de fin d'études, INSA Lyon, 37 pp.
23. Tarradellas, J., Diercxsens, P., and Bouché, M.B. (1982) Methods of extraction and analysis of PCBs from earthworms, *Int. J. Environ. Analyt. Chem.* **13**, 55-67.
24. Abdul Rida, A.M.M. and Bouché, M.B. (1995) The eradication of an earthworm genus by heavy metals in Southern France, *Appl. Soil Ecol.* **2**, 45-52.
25. Ireland, M.P. (1983) Heavy metal uptake and distribution in earthworms, in J.E. Satchell (ed.), *Earthworm Ecology, from Darwin to Vermiculture*, Chapman and Hall, London, pp. 247-265.
26. Ma, W.C. (1987) Heavy metal accumulation in the mole, *Talpa europea*, and earthworms as an indicator of metal bioavailability in terrestrial environments, *Bull. Environ. Contam. Toxicol.* **39**, 933-938.
27. Morgan, J.E. and Morgan, A.J. (1992) Heavy metal concentrations in the tissues, ingesta and faeces of ecophysiologically different earthworm species, *Soil Biol. Biochem.* **24**, 1691-1697.
28. Marquenie, J.M. and Simmers, J.W. (1984) Bioavailability of heavy metals, PCB and PCA components to the earthworm *Eisenia foetida*, in *Environmental Contamination*, UNEP, London.
29. Ebing, W., Pflugmacher, J., and Haque, A. (1984) Der Regenwurm als Schlusselorganismus zur messung der Bodenbelastung mit organischen Fremdchemikalien, *Berichte über Landwirtschaft* **62**, 222-255.
30. Edwards, P. and Kreis, B. (1985) *Présence et cheminement de micropolluants organiques dans les peuplements lombricienns des*

agroécosystèmes, Rapport École polytech. fédér. Lausanne, 73 pp.

31. Walsh, P., El Adlouni, C., Nadeau, D., Fournier, M., Coderre, D., and Poirier, G.G. (in press). DNA adducts in earthworms exposed to contaminated soils, in *Proc. 5th Intern. Symp. Earthworm Ecology*, Columbus, Ohio, USA.
32. Lee, K.E. (1984) *Earthworms: Their Ecology and Relationships with Soils and Land Use*, Academic Press, London.
33. Edwards, C.A. and Lofty, J.R. (1972) *Biology of Earthworms*, Chapman and Hall, London.
34. Muller, P.E. (1889, translated) Recherche sur les formes naturelles de l'humus, *Ann. Sci. Agron.* **1**, 1-143.
35. Bornebusch, C.H. (1930) The fauna of forest soil, *Pet. Forstigge Forsogsvasen* **11**, 1-158.
36. Granval, Ph. and Aliaga, R. (1988) Analyse critique des connaissances sur les prédateurs des lombriciens, *Gibier et Faune sauvage* **5**, 71-94.
37. Abdul Rida, A.M.M. and Bouché, M.B. (1995) Earthworm contribution to ecotoxicological assessments, *Acta Zool. Fennica* **196**, 307-310.
38. Hadjibiros, K. (1977) *Simulation d'une système prédateur-proie à partir d'un modèle matriciel*, Rapport DGRST, action concertée "Système édaphique: analyse, synthèse et gestion", 146 pp.
39. Samih, M. (1988) *Quelques problèmes spécifiques à la modèlisation non linéaire: applications à des problèmes écologiques*, Thèse doct. Mathématiques pures et appliquées, USTL Montpellier, 105 pp.
40. Bouché, M.B. (1984) Écotoxicologie des lombriciens. I. Écotoxicité contrôlée, *Oecol. applic.* **5**, 271-287.
41. Savanne, D. (in press) French research programme on methods for investigating the ecotoxicity of contaminated soils and waste, in *Proc. 5th KfK-TNO Conference on Contaminated Soils*, Maastricht, 1995.
42. Bouché, M.B. (1992) Earthworm species and ecotoxicological studies, in P.W. Greigh-Smith, H. Becker, P.J. Edwards, and F. Heimbach (eds.), *Ecotoxicology of Earthworms*, Intercept, Andover, pp. 20-35.
43. Donker, M.H., Eijsackers, H., and Heimbach, F. (eds.) (1994) *Ecotoxicology of Soil Organisms*, Lewis Publ., Boca Raton, 470 pp.

INSECTS IN BIOINDICATION OF SOIL POLLUTION

R.O. BUTOVSKY
All-Russian Research Institute for Nature Protection
Sadki-Znamenskoje
113628 Moscow
Russian Federation

Abstract

An analysis of the scientific literature shows that soil insects are characterized by a high biodiversity in various respects (species diversity, diversity of life-forms, morphological diversity, *etc.*). They represent a unique group, united by the peculiarities of their morphology, ontogeny and ecology. The set of responses to anthropogenous influences seems to be universal for all groups of soil insects, which could be explained by their common origin and similar pathways of input of pollutants into the environment. Among the various trophic groups the most sensitive are saprophageous insects and non-specialised predators, the most tolerant groups seem to be the phytophageous insects, parasites and specialised predators.

1. Introduction

Interrelations of man with insects have a dualistic nature, from the viewpoint of man. On the one hand man tried to fight insect pests, that destroyed his crops and food supplies, and by doing so, man pollutes the environment with enormous amount of pesticides; on the other hand man has finally recognized the significance and importance of insects as a necessary component of biodiversity on earth and even more or less successfully tried to save some groups from extinction.

To keep the right balance between purposeful and purposeless chemical elimination of insects by man and the conservation of insect biodiversity, we need: 1) to study insect responses to different chemical pollutants, 2) to study the possibilities of adaptation to different chemical pollutants, 3) to predict scenarios of changes in insect communities exposed to pollutants, and 4) to select universal responses that could be used for bioindication purposes.

N. M. van Straalen and D. A. Krivolutsky (eds.), Bioindicator Systems for Soil Pollution, 155–163.

The use of insects for bioindication is not new. Moreover, many groups of soil insects have been used "traditionally" for assessment of environmental quality and side-effects of pesticides, for example, Collembola, ground-beetles, and rove-beetles.

2. Biodiversity of soil insects

The biodiversity of insects is well-known and this may be one of the main reasons for their great flexibility towards various environmental stresses. In one of the best studied faunas of the world, the British, the most numerous (about 20,000 species) are insects, followed by spiders and mites (3000 species), and other invertebrates, such as myriapods and isopods (about 3600 species) [1].

Being a successful invertebrate group, with numerous adaptations on different levels, insects take a relatively high part of the biomass and species diversity in the soil fauna. In spite of the fact that many groups of insects are connected with soil and litter in their early life-stages (like Diptera larvae, wireworms, cutworms, *etc.*), only a few groups of insects are supposed to be "real" soil inhabitants: Diplura, Protura, Collembola, Thysanura, ants, and some families of Coleoptera (among which the Carabidae and Staphylinidae are best known).

According to some evaluations the approximate number of nematode species in soil was estimated as 500 - 1000 thousand; for other groups species numbers were estimated as follows: oligochaetes: 1200 species, crustaceans: a few hundred species, oribatids: 7000, myriapods: 3000, and diplopods: 7000. Species richness of Collembola was estimated to be 10,000 to 20,000 species, while 1200 species were estimated for soil-dwelling bugs, 100 for Embioptera, 50 for gryllotalpids, 3500 for cockroaches, 40,000 for ants, and 27,000 for staphylinids [2].

These theoretical estimations may be compared with real data on species diversity for some relatively well studied soil groups in the former USSR and in the Moscow region (Table 1). These comparisons clearly show that insect species richness in general is higher than biodiversity of any other groups of invertebrates.

3. Insects as a unique group

As an object of study and as a part of wildlife, insects represent a unique group, possessing a set of peculiarities that stem from their morphology, ontogeny and ecology. Following G.N. Gornostaev [17], we may note five important aspects: body-size, metamorphosis, dispersal, phenology and turnover.

TABLE 1. Comparative species diversity of soil animals in the former USSR and the Moscow region

Taxa	Number of species	Region	Ref.
Collembola	300	USSR	[3]
Collembola	150	Moscow	[4]
Cockroaches	50	USSR	[3]
Carabids	2500	USSR	[5]
Staphylinids	385	Moscow	[6]
Nematodes	500	Moscow	[7]
Oribatei	340	Moscow	[8]
Gamasidae	204	Moscow	[9]
Earthworms	97	USSR	[10]
Earthworms	11	Moscow	[11]
Slugs	102	USSR	[12]
Gastropoda, Pulmonata, Geophila	170	USSR	[13]
Helicoidea	166	USSR	[14]
Lithobiomorpha	126	USSR	[15]
Diplopoda	190	USSR	[16]
Julidae	96	USSR	[16]
Polydesmidae	35	USSR	[16]

Body size. Many species of insects are connected with small and sometimes very special habitats during their development. According to a general rule [18] the smaller an animal is, the shorter is the duration of unfavourable periods; in addition, the less time an animal needs to complete development and reproduction, the more possibilities it has to discover the optimal set of environmental conditions.

Metamorphosis. Development through moultings and metamorphosis is connected with some basic differences in the morphology and ecology of the different stages. During ontogenesis one can observe changes of ecological niches and larval habitats; moreover, even imagines of the same species can be completely different (*e.g.* sexual dimorphism).

Dispersal. The ability to fly and the tendency to disperse or migrate is another remarkable aspect of the life of an insect.

Phenology. A marked annual or perennial pattern of abundance is characteristic for many species. It is important, that during the decrease of abundance uniform large populations are subdivided into small isolated groups.

Turnover. A high natural mortality is compensated by high fecundity of females and a rapid succession of generations.

4. Scenarios of insect responses to chemical pollutants

It is important to stress that the behaviour of different pollutants in the environment and their influences on insect communities and ecosystems show many similarities, irrespective of their origin. According to Krivolutskii [2] pathways by which pollutants enter the environment are similar for all chemicals. The main sources are atmospheric deposition,

TABLE. 2. Array of responses of soil insects under anthropogenic influences

Level	Responses
Community	Trophic structure Secondary productivity Functional structure Community structure or relative abundance of species (genera, families, orders etc.) Species diversity
Population	Population density Intrapopulation diversity Size-mass structure Sex ratio Age structure
Organism	Chemical composition Morphological abnormalities Real and potential fecundity Differential lethality in ontogenesis Differential longevity in ontogenesis Activity of fermentative systems

input into reservoirs, and soil and plant pollution by irrigated water. When pollutants are deposited on plant and soil surfaces, their physico-chemical appearances are similar too. They are mainly present as dispersed aerosoles or atmospheric precipitates, which are composed of different pollutants, although sometimes chemical pollutants are present in the gaseous phase. Another aspect of similarity is due to the fact that the fate of different pollutants is determined by the same general laws of partitioning and distribution in soils and on plant surfaces.

Finally, a close correlation exists between species diversity and different types of chemical pollutants. It means, that if one species is vulnerable to one type of pollutant, very often it is also vulnerable to other types of pollutants. That is why all invertebrate groups show more or less the same set of responses to different types of chemical pollution. This implies that we can expect: the same set of responses to anthropogenic influences and a similar pattern of change in soil invertebrate communities. These aspects are enumerated in Tables 2 and 3.

We suppose that at present the community level is the most informative level to be used in bioindication. When we move from community to the organism level, the bioindicator value of the responses decreases. Probably this is related to an increase of the number of factors that influence the expression of a response and to the correspondence between a particular response and the level of anthropogenic influence. The scenarios of changes of soil invertebrate communities also seem to be common, independent of the type of influence (Table 3).

5. Conclusions

When insects are used for bioindication of soil pollution, it is important to realize the following aspects: 1) soil insects have a high biodiversity in the general sense of the term (species richness, diversity of life-forms, morphological diversity, *etc.*), 2) soil insects are a unique group, united by the peculiarities of their morphology, ontogeny and ecology, 3) the set of responses to anthropogenous influences is similar for many groups of soil insects, which can be explained by their common origin and common pathways by which pollutants enter the environment, 4) among the various trophic groups the most sensitive ones are saprophageous species and non-specialised predators, the most tolerant seem to be the phytophageous species, parasites and specialised predators.

TABLE 3. The main patterns of change observed in soil invertebrate communities exposed to chemical pollution

Impact	Scenario	Ref.
Motorway influences	Decrease of abundance of saprophageous insects (Diptera larvae), non-specialised predators (carabids, staphylinids), increase in abundance of specialised predators, parasites, phytophageous insects, root-eaters (wireworms, curculionids, scarabaeids), and necrophageous insects Decrease in abundance of saprophageous species in microarthropod communities. Increase of species diversity under moderate stress.	[19]
Heavy metal pollution	Decrease of abundance and species diversity of sensitive groups. The ability to accumulate and store heavy metals usually decreases in the following sequence: annelids, terrestrial crustaceans (woodlice and amphipods), arachnids (spiders and mites), hemimetabolic insects, holometabolic insects. In roadside ecosystems polluted with heavy metals, the most sensitive component can be carabids and staphylinids. High heavy metal concentrations lead to a decrease in reproductive potential and a decrease of offspring size.	[20]
Pollution by rare elements	Decrease of abundance of mesofauna in general (all insect larvae, imagoes of beetles, staphylinids, spiders, enchytraeids), saprophageous species in general. Decrease of species diversity (staphylinids, Diptera larvae, carabids)	[2]
Neutron and gamma radiation	Decrease of abundance of beetles, insect larvae, enchytraeids. Increase of activity of active dispersers (adult beetles). Decrease of species diversity in deep soil layers. Maintenance of Collembola densities at high levels.	[2]
Pollution by sulphur containing substances	Decrease of the total abundance of soil animals after 10 days following a catastrophic emission, persistence of spiders, enchytraeids, earthworms, diplopods, wireworms and curculionids. After 30 days the abundance of enchytraeids and	[21]

TABLE 3, continued

	wireworms decreased and abundance of predators increased. After 90 days the total abundance of all trophic groups of animals decreased. After repeated emissions after the total abundance of soil animals increased after 30 days. After 90 days the community structure has changed: the abundance of diplopods, enchytraeids and Diptera larvae increased, the abundance of earthworms decreased.	
Fluor pollution	Decrease of abundance and biomass of carabids and staphylinids. Increase of abundance and biomass of plant-sucking groups. Decrease of number and average volume of ant nests, average mass of individual workers and average size of nests. Decrease of abundance of litter inhabitants and saprophageous species.	[22]
Dust pollution (cement and lime dust)	Decrease of abundance of Diptera larvae, Collembola, and carabids. The most pronounced changes were observed among invertebrates of litter and top soil layers.	[23] [24]
Crude oil pollution	Mass elimination of soil life in intensively polluted areas. The most pronounced effects can be observed during the first days after a spill. Insects are a very sensitive group in comparison with Protozoa and nematodes.	[25]
Oil industry sewage	Decrease of abundance among soil organisms, particularly Collembola. Decrease of species diversity, changes in dominance structure (*e.g.* in carabid communities). Elimination of arthropods in the top 5 cm soil layer. Increase of the concentration of chloride leads to a decrease in abundance of wireworms and earthworms.	[25]
Pesticide application	Decrease of abundance and species diversity of soil organisms. The most sensitive are the benificials: staphylinids and carabids. Some phytophageous and saprophageous species increased their abundance; development of genetic resistance within a few generations was observed in some groups of pest insects.	[19]

6. References

1. Kirby P. (1992) *Habitat Management for Invertebrates. A Practical Handbook*, Royal Society for the Protection of Birds Publ., London.
2. Krivolutskii, D.A. (1994) *Soil Fauna in Ecological Control*, Nauka, Moscow (in Russian).
3. *Key to Insects of the European part of the USSR, V 1.* (1964) Nauka, Moscow (in Russian).
4. Solntseva, E.L., Babenko, A.B., Kuznetsova, N.A., Nadtochi, S.E., and Uvarov, A.V. (1982) Collemboles (Collembola) of Moscow region, in *Soil invertebrates of Moscow region*, Nauka, Moscow, pp. 97-107 (in Russian).
5. Kryzanovskii, O.L. (1983) *Beetles of suborder Adephaga: Families Rhysodidae, Trachypachidae; family Carabidae (Introduction, review of the USSR fauna)*, Nauka, Leningrad (in Russian).
6. Tihomirova, A.L.(1982) Fauna and ecology of staphylinids (Coleoptera, Staphylinidae) of Moscow region, in *Soil Invertebrates of Moscow Region*, Nauka, Moscow, pp. 201-223 (in Russian).
7. Gubina,V.G. (1982) On nematological studies in Moscow region, in *Soil Invertebrates of Moscow Region*, Nauka, Moscow, pp. 41-46 (in Russian).
8. Krivolutskii, D.A., Chugunova, M.N., Gordeeva, E.V., and Tarba, Z.M. (1982) Fauna of oribatids (Acariformes, Oribatei) of Moscow and adjacent regions, in *Soil Invertebrates of Moscow Region*, Nauka, Moscow, pp. 55-72 (in Russian).
9. Petrova, A.D. (1982) On fauna of soil-dwelling gamasidae mites (Parasitiformes, Mesostigmata) of Moscow region, in *Soil Invertebrates of Moscow region*, Nauka, Moscow, pp. 77-84 (in Russian).
10. Perel, T.S. (1979) *Range and Regularities in the Distribution of Earthworms of the USSR Fauna* (with keys to Lumbricidae and other Megadrili), Nauka, Moscow (in Russian).
11. Matveeva, V.G. and Perel, T.S. (1982) Earthworms of the family Lumbricidae of Moscow region, in *Soil Invertebrates of Moscow Region*, Nauka, Moscow, pp. 133-144 (in Russian).
12. Likharev, I.M. and Victor, A.I. (1980) Slugs of USSR fauna and adjacent countries (Gastropoda terrestria nuda), in *Fauna of the USSR.* Mollusca. V. III. issue 5, Nauka, Leningrad (in Russian).

13. Schileyko, A.A. (1984) *Terrestrial Snails of Family Pupillina (Gastropoda, Pulmonata, Geophila)*, Nauka, Leningrad (in Russian).
14. Schileyko, A.A. (1978) Terrestrial snails of family Helicoidea, in *Fauna of the USSR.* Mollusca. V. III. issue 6, Nauka, Leningrad (in Russian).
15. Zalesskaya, N.T. (1978) *Key to Myriapods (Chilopoda, Lithobiomorpha) of the USSR Fauna*, Nauka, Moscow (in Russian).
16. Lokshina, I.E. (1969) *Key to Diplopoda of Plane Part of the European part of the USSR*, Nauka, Moscow (in Russian).
17. Gornostaev, G.N. (1986) Problems of conservation of extincting insects, in *Totals of Science and Technics*, VINITI Publ., Moscow (in Russian).
18. Ghilarov, M.S. (1944) The ratio between the sizes and abundances in soil invertebrates, *Doklady Akademii Nauk. USSR* **43** (6), 283-285 (in Russian).
19. Butovsky, R.O. (1991) *The Conservation of Beneficial Insects in Conditions of Chemical Environmental Pollution.* VNIITEI agroprom Publ., Moscow (in Russian).
20. Butovsky, R.O. (1993) Heavy metals in terrestrial arthropods. 2. Insecta, *Agrohimija* **7**, 113-124 (in Russian).
21. Ryabinin, N.A., Ganin, G.N., and Pankov, A.N. (1988) Influence of sulphur acid wastes on assemblages of soil invertebrates, *Ecologia* **6**, 29-37 (in Russian).
22. Kataev, O.A., Golutvin, G.I., and Selihovkin, A.V. (1983) Changes in arthropod communities of forest biogeocenoses under the atmosphere pollution, *Entomol. Review* **62** (1), 33-41 (in Russian).
23. Kozlov, M.V. (1990) The influence of anthropogenous factors on the populations of terrestrial insects, *Totals of Science and Technics.* V. 13. VINITI Publ., Moscow (in Russian).
24. Melecis, V.P. (1985) Bioindicative significance of collemboles under pollution of birch forest by calcium-containing dust, in *Pollution of Environment by Calcium-containing Dust*, Zinatne Publ., Riga, pp. 149-209 (in Russian).
25. Artemyeva, T.I. (1989) *Assemblages of Soil Animals and Problems of Recultivation of Technogenous Territories*, Nauka, Moscow (in Russian).

SOIL SPIDERS AND BIOINDICATION

J.-P. MAELFAIT

Instiute of Nature Conservation
Kiewitdreef 5
B-3500 Hasselt
Belgium

1. Introduction

The purpose of this contribution is to illustrate the possiblities of spiders in ecotoxicological field studies. First (paragraph 2), I give a general introduction to spiders, stressing the characteristics they have which are of importance for ecotoxicological research. Secondly, I summarise what is known about heavy metal contamination and accumulation in spiders (paragraph 3). Thirdly, I will describe the methods we normally use to sample spider communities, that is the pitfall trapping method. I will try to show that with that sampling technique we can study: (1) The life and activity cycle of ground-dwelling spiders; this has implications for the bio-availability of pollutants for higher trophic levels (paragraph 4.2) and (2) the environmental factors, like for instance pollution stress, affecting spider communities (paragraph 4.3). Fourthly and finally I present some results of an arthropod sampling campaign we made on the banks of a polluted river in Belgium (paragraph 5).

2. Spiders: a general introduction

Figure 1 shows a schematic drawing of a spider. Being arthropods, spiders have an exoskeleton and tubular legs with a number of joints. Together with ten other orders, like the scorpions, the pseudoscorpions, the mites and the ticks and the harvestmen, spiders belong to the arachnids. Unlike insects, arachnids have four pairs of walking legs; they do not possess antennae or wings and have simple rather than compound eyes. Spiders differ from other arachnids because their body consists of two separate parts: the cephalothorax and the abdomen or opisthosoma. The two parts are connected by a narrow tube, the so-called pedicel. Apart from the four pairs of walking legs, they have a a pair of pedipalps and a pair of chelicerae. The pedipalps are in the adult males enlarged into copulatory organs. The basal part of each chelicera carries a movable tooth, known as the fang. The chelicerae have ducts from a poison gland opening at the top of the fangs. When a prey has been captured

N. M. van Straalen and D. A. Krivolutsky (eds.), Bioindicator Systems for Soil Pollution, 165–178.

it is subdued by an injection of venom through the piercing fangs. Apart from that movable tooth, most spider families also have chitinous protrusions on the chelicerae, the so-called cheliceral teeth.

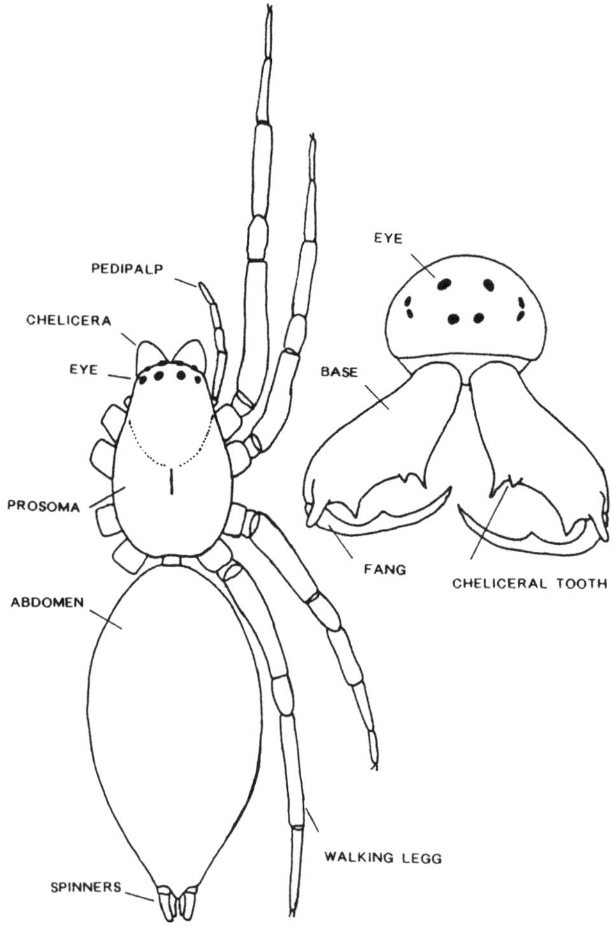

Figure 1. Schematic drawing of a spider.

More than 50,000 species of spiders have been described exhibiting a wide variety of shape and of life strategies. They all produce silk in abdominal glands, which open in the so-called spinners. By means of that silk most spiders construct webs to catch their prey, mostly flying and crawling insects. Quite a lot of species do not make webs but are free-living hunters which only use silk to construct retreats in which they spend either night or day and to make egg sacs. Silk is also used as a help in moving through the vegetation or as a means to travel through the air, the so-called ballooning. A silk thread is trailed after walking animals; it can be used as a safety line when dropping out of sight of predators. There are more than 700 spider species living in northern Europe varying in size from about one millimeter to a few

centimeters. With one exception (the water spider, *Argyroneta aquatica*) spiders are all terrestrial. The life cycle of a spider can be summmarised as follows:
-eggs are laid in a clutch which is covered to a greater or lesser extent with silk
-the young emerging from the egg sac, the spiderlings, are small versions of their parents; they only have relatively shorter legs. Development is gradual: having an exoskeleton, spiders grow by a series of moults. The number of these moults is about five in small species and about ten in large ones.
Spiders are all predators. They are not able to ingest solid food particles. Therefore, digestion must be external. Digestive fluids are poured onto or into the immobilised prey. Species with cheliceral teeth mash their prey to a ball of inedible parts, others suck the digested tissues through the pore made by the fangs, leaving an externally undamaged prey. In that way spiders feed themselves solely with the soft tisssues of their prey. Because most pollutants are stored in these tissues, spiders ingest higher concentrations than the concentrations in the whole body of the prey items. Spiders mostly feed on all kinds of insects and other arthropods. Cannibalism is also widespread, mostly because older stages eat younger ones; occassionally scavenging may occur. On their turn, spiders form important prey for birds, reptiles, amphibians and small mammals. They may, for that reason, be of quite some importance in the transfer of pollutants from the lower to the higher levels of the food chain. Most spiders living in our regions have a life cycle of about one year; some smaller species have two to three generations per year. In bioindicator research, short life cyles are generally considered to be an advantage, because short-lived populations are able to react more quickly (through changes in numbers, for example) to changing environmental conditions.

3. Spiders and heavy metals: a short review of the literature

Very roughly, heavy metal accumulation in invertebrates can be summarised as follows. Woodlice, snails and earthworms accumulate significant amounts of metals, whereas insects are able to regulate concentrations to relatively low levels. What about spiders: do they and, if so, how do they fit in that generalisation? To be able to answer that question, we summarise hereafter the results of a few studies.
Although not really bearing on ground living spiders, we will first consider the earlier work of Clausen [1]. This author studied the possibilities of spiders as indicators for atmospheric lead pollution in Denmark. He concluded that the lead concentration in the web-building spiders *Nuctenea umbraticus* was as good a measure of atmospheric lead pollution as the concentration of lead found in a lichen. About 25% of the lead was deposited at the surface of the hairy body of the spider. The lead within the body originated from ingested prey but also from eating the old web prior to producing a new one. As has been found for several other groups; Clausen's results showed that considerable differences in heavy metal concentrations are exhibited by species which are closely related taxonomically and which occur at the same site.
Hunter and co-workers [6] found in pooled pitfall samples of Linyphiids, small web-

building spiders, as concentrations, expressed as ppm on whole body dry weight: (1) close to a copper refinery: 1020 for copper and 87 for cadmium and (2) in their control site: 89 for copper and 2.4 for cadmium. For pooled samples of Lycosids (a family of hunting spiders), these figures were respectively: (1) 887 for copper and 102 for cadmium and (2) 58 for copper and 2.6 for cadmium. A large seasonal variation was observed in these concentrations. This was attributed to the presence in the samples of different species and to the recruitment of lesser contaminated juveniles. They found that in spiders the mean concentration of copper was two to three times higher than in predatory beetles, for cadmium this was 7 to 8 times higher. Only isopods had higher concentrations, but these isopods were strongly reduced in abundancies, so that spiders were the major source for cadmium accumulation in insectivorous small mammals, like the common shrew (*Sorex araneus*). Spiders made up 30 to 70% of the diet of that mammal. This percentage was especially high during winter, when the abundance and availability of other prey items, such as insect larvae and adult beeetles were very low.

Van Straalen and Van Wensem [10] found near a zinc smelting work in the Netherlands, of all investigated animals the highest cadmium concentration occurred in the spider *Centromerus sylvaticus*, about 20 ppm.

For the ground-living and woodlice-eating spider *Dysdera crocata* it was shown by experiments done by Hopkin and co-worker [3,4] that this species was able to prevent zinc, cadmium and lead from being assimilated even if these animals were fed with woodlice, containing very large amounts of these heavy metals. These unwanted metals are stored in granules in the hepatopancreas of the spider; they can be voided with the faeces. Although, indeed kept below certain levels, the concentrations in the spider body of these metals are still very high: about 1200 ppm for zinc, about 1100 ppm for copper and about 40 ppm for cadmium.

In experiments with spiders fed on fruit flies, contaminated with cadmium, Clausen [2] showed that the spider *Steatoda bipunctata* behaved as a regulator for lead . The same species was however an accumulator for cadmium. The levels of cadmium in fruit flies were 5 to 7 times biomagnified in the spiders, reaching levels of 50 ppm and more; these are concentrations which could be toxic for for instance birds predating on these spiders.

From this short review of the literature we can conclude that heavy metal concentrations in spiders can be very high. This is certainly true for the non-essential heavy metal cadmium. Even in spider species which seem to be able to regulate the concentrations of that element, it reaches levels which could be toxic for spider-eating predators.

4. Pitfall sampling of spiders

4.1. THE METHOD

To investigate variation in ground living spider and carabid communities, we normally use pitfall trapping as a sampling method. The method requires only a relatively small investment in time and equipment. For most terrestrial habitat types, with the exception perhaps of riparian habitats, the method is much more effective than quadrat sampling and hand-collecting. If project budgets allow, however, both other methods can give valuable additional information. The traps we use are very cheap; they are glass jars of half a liter, normally used as jam jars. The traps are usually operative during a complete year cycle. They are emptied and installed again at regular time intervals, usually every fortnight. Dependent on the expected evaporation and rainfall for the period and site of operation, they are filled for one third to two thirds with a 10% of formalin solution. Three to six traps are used per sampling site. Because different species are active during different seasons, trapping should be done all year round to obtain a more or less complete picture of the biodiversity of the studied habitats. We will now discuss some of the more important applications for which the traps can be used. The model we work with to interpret the captures obtained by pitfall traps is as follows [7]. The Capture, in other words the number of animals of a particular species (sp) trapped, during a particular period of time (pe), with a pitfall trap, placed at a particular site (si) is dependent of, is proportional to the activity of that species, during that period at that site and also proportional to the Density of that species, during that period at that site:

$$C_{sp,pe,si} = cte \, . \, A_{sp,pe,si} \, . \, D_{sp,pe,si}$$

Accepting that model, implies that we:
1) are comparing Densities by comparing Captures if we can accept that Activities are constant for the compared situations
2) are comparing Activities by comparing Captures if we can accept Densities to be constant (or nearly so: see further).

4.2. SEASONAL ACTIVITY PATTERNS OF POPULATIONS

A first application of the above given deductions from the model is the interpretation of the constant patterns we observe in the distribution of the captures of a particular species during a year cycle. For very meany ground-living spiders and other epedaphic arthropods we see that the captures made during periods spaced over the year give a picture as shown in Figure 2. The distribution patterns of captures made during a year cycle are constant from year to year. During most of the year, very few, mostly juvenile and subadult individuals are caught; during a short period a large number, mostly adult specimens are caught. These distribution patterns can differ to a lesser or greater extent between the two sexes. The distribution patterns are constant within and are thus typical for the species, at least over a wide geographical range. How can we interpret these distribution patterns? We also observe that during that period of large captures of adults, eggs are laid and young stages of juveniles appear. Now we know that for most species density at birth is very high but drops very rapidly, as shown in Figure 2. When we compare that graph with the graph of the temporal distribution of captures we can conclude that density changes are certainly not responsible for the seasonal pitfall capture distributions.

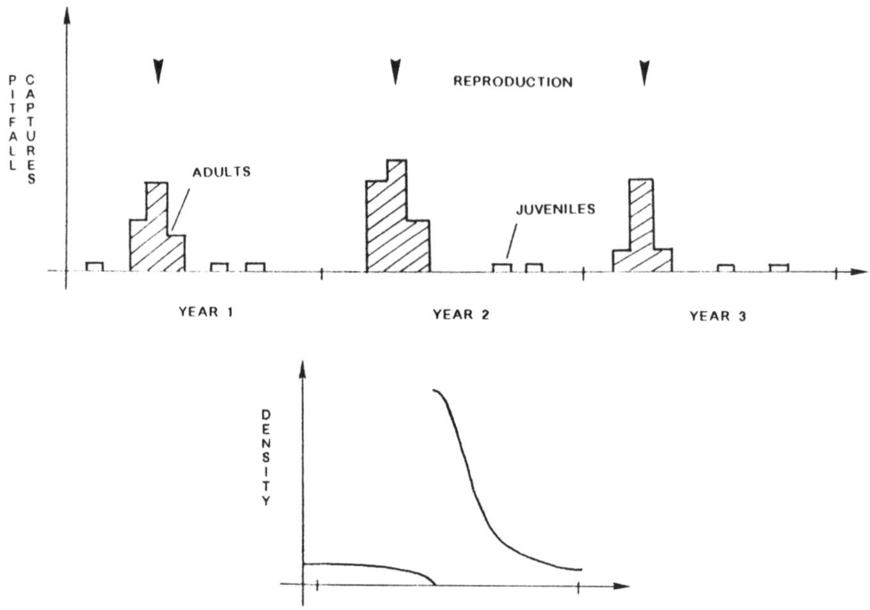

Figure 2. Pitfall capture distribution during year cycle, compared with density variation (see text).

The temporal distribution of the captures must, therefore, be largely influenced by variation in the other parameter determining the capture rate, in other words by variation in the ground surface activity of the involved species. Young animals, although they are very abundant, are hardly caught by pitfall traps because of their very low ground surface activity. They live hidden between the vegetation or in the litter layer and only very rarely move about on the ground surface. The reverse applies to the adults: low density, but high bottom surface activity. The difference in bottom surface activity between juveniles and adults is so pronounced because the adults' activity is highly related to reproduction. Adult males are so active because they are in search for females to inseminate. Females are more active because they are in search for suitable places for hiding their egg sacs and young or for basking their egg sacs in the sun, like lycosid spiders do. What can be done with that knowledge about the behaviour of ground-living spiders within the framework of contamination by and accumulation of pollutants? I think, it can serve as an example to illustrate that bio-availability is not only a matter of concentrations within a given prey species but also of the availability as a prey item to higher trophic levels. Young spiders do not run about, they are mostly hidden. One would therefore expect them to become the prey of predators that activily search between vegetation and litter, like for instance insect-eating mammals, predatory beetles and other spiders. Adult spiders, on the other hand, which may build up larger concentrations of pollutants, are because of their active behaviour much more conspicuous for, and prone to be caught by sight predators, like for example birds, frogs and lizards. Analysis of the life and activity cycle and the availabity for predators is one application of pitfall trapping.

4.3. ENVIRONMENTAL FACTORS AFFECTING SPIDERS

Another purpose for which we use pitfall trapping of spiders are studies concerning the environmental factors that influence the composition of spider communities. In other words studies concerning the environmental factors that influence the distribution and abundance of spider species. Because, as we saw, pitfall captures not only depend on abundance but also on activity, we must have good reasons for assuming that for the situations we compare, activities are not different. That this condition would be fulfilled when comparing different species between each other is highly improbable. We therefore do not use pitfall captures to compare the density of one species to the other. A situation, for which we can accept the condition of equal activities to be fulfilled, is when using pitfall captures made during a complete year cycle on several sites in order to study the spatial distribution pattern of a species. With such kind of pitfall captures we can analyse the habitat choice of a species, we can investigate which environmental factors influence the occurrence and abundance of a species. By using indirect and direct ordination techniques this analysis can be done for an assemblage of species and for several environmental factors at the same time. When we apply these multivariate analyses we only use the most abundantly caught species and we give each species equal weight. Now I will try to summmarise some examples of that kind of synecological investigations we made within the framework of nature conservation rersearch.
As a first example I can mention the results that we obtained in a comparison of several woodland stands [9]. In our region, like in most other European regions, almost all forests have been managed by foresters. That implies that our remaining woodlands are split up in small stands consisting of one or very few species of trees, often of non-indigenous origin and each stand with its own management regime. This leads of course to a very artificial, a man-made habitat structure. One of the aims of nature conservation in our region is to obtain again more natural forests. Within that context we sampled several of these forest stands varying in the composition of their tree layer, their soil and litter characteristics and in their hydrological regime. For the series of stands we sampled we found out that the seasonal variation in groundwater level, the nutrient level of the upwelling water and the soil texture were the most important factors determining the variation in the composition of the ground-living spider communities. Of much less importance was the composition of the tree layer. To our opinion, this result gives hope for the future: in spite of artificial tree layers with a low diversity, there still remains a high biodiversity in the ground-living spider communities of these forests. These communities will develop themselves even more when in forest management much more opportunities will be given to spontaneous developments.

A second example concerns what we found to be the more important habitat features influencing the ground-living spider communities of heathland [8]. Heathland is a plagioclimax vegetation of dwarf shrubs, with a height of 0.5 to 1 m. This habitat type resulted from the traditional exploitation of woodland habitats which occurred on poor sandy soils. Nowadays, nature conservation wants to keep these typical semi-natural habitats. To prevent they evolve to woodland they have to be managed. This can be done by mowing, burning, grazing or by cutting young invading tree seedlings. These different management techniques lead to differently structured vegetations. In order to assess the effects of these differences on the ground-living arthropod communities, we sampled a number of dry heathland sites differing from each other in previous management and thus in vegetation structure. We found that the most important gradient of variation could be associated with the age of the heathland vegetation. Recently managed sites with young stands of heath had a relatively poor spider fauna mostly composed of ubiquistic species, while old heathland had a species rich spider community with several species which are quite rare in our region. This could be explained by the fine grained mosaic in old heathland of open patches with a thin cover of lichens and mosses intermingled with patches of old heath plants with a thick litter layer. This means that in old heathland open sunny microhabitats for an optimal reproduction and growth during summer occur very closely to microhabitats which give protection for surviving harsh winter conditions.

From the above summarised studies and from comparable investigations in coastal dunes and in grasslands, we can conclude that variation in the diversity and the composition of spider communities is to a large extent determined by the structure of the field layer, the litter layer and some other edaphic factors. This makes spiders good candidates for what Hopkin [5] recently called 'in situ biological monitoring of pollution on presence or absence of species from a site, the so-called community effects of pollution'. Indeed pollution often causes changes in vegetation and litter structure and spiders are quite sensitive to such changes. Of course, more direct effects might also occur.

5. HEAVY METAL CONTAMINATION IN LITTORAL SPIDERS

The river Scheldt rises in northern France and after some 350 kilometers it discharges in the North Sea, in The Netherlands (Figure 3). The river is tidal up to Ghent. Sea water enters and leaves the estuary twice a day with the tides. There is only salty water penetration till the city center of Antwerp, but, due to the back-up effect, the water level fluctuates up to Ghent. At that city there is still a water level difference between high and low tide of two meters. Along the river side of the dikes, at several places marshes occur. These marshes are tidal. Their lower parts are mud flats which are flooded twice a day. Their higher parts are vegetated either by grasses and forbs or by willow scrubs; these are only flooded during spring tide, that is twice a month. In twenty five of these marshes we sampled the non-woody vegetations by means of a time standardised effort of hand-collecting. On the whole more than 50 species of woodlice, talitrid amphiphods, carabid beeetles and spiders were collected. In these frequently flooded habitats, earthworms are only very exceptionally encountered. Multivariate analysis of the data revealed that the salinity of the water of the river was the most important factor determining the composition of the arthropod communities.

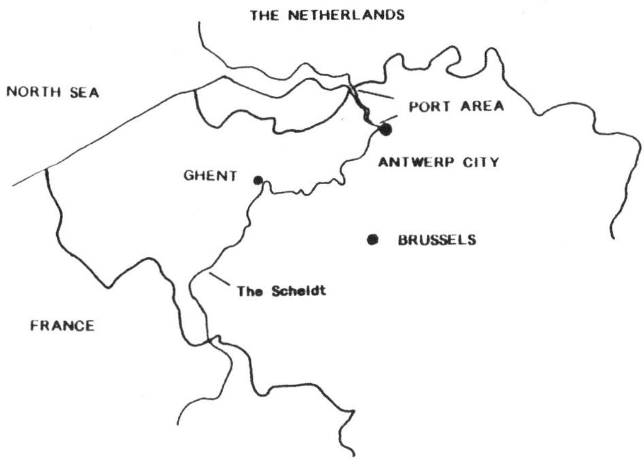

Figure 3. The Scheldt basin.

Salt and brackish marsh communities are present uptill the city center of Antwerp; higher up we find communities built up of species bound to freshwater conditions. Good indicators for that rather abrupt change are the two sister species of terrestrial amphipods; *Orchestia gammarella* and *O. cavimana.* The first one occurs in pure populations up to the city of Antwerp; the other in pure populations upstream of that city. Due to the rapid economic development of our regions after World War II and that without enough attention for the ecological consequences of that growth, all rivers in the northern part of Belgium are still heavily polluted. The underwater soils are contaminated by industrial and agricultural discharges. It seemed therefore quite interesting to look for the heavy metal concentrations, present in the animals we collected on the littoral soils along the Scheldt. The heavy metal determinations were in part done by Dr. R. Butovsky (All-Russian Reasearch Institute for Nature Protection, Moscow) during a two month stay at our Institute with a grant of the European Science Foundation. I will not treat here all results we obtained, but I will only draw your attention to some figures which are of some more general interest.

In Table 1 copper, zinc and cadmium concentrations in ppm of whole body dry weight are given for six detritivorous and three spider species occurring in the brackish marsh 'Het Galgeschoor' situated in the port of Antwerp. All animals collected at that site contained cadmium. The concentrations for that heavy metal are quite high in the spiders. The copper and zinc concentrations of the adults of the spider *Pardosa purbeckensis* are more than twice as high as in the juveniles, while the cadmium concentrations are the same. In *Trochosa ruricola* copper and zinc concentrations are very much alike while the cadmium concentration in juveniles is more than twice higher than in adults.

TABLE 1. Copper, zinc and cadmium concentrations five isopod species, a terrestrial amphipod and two spider species in the brackisch tidal marsh 'Het Galgeschoor'.

Species	Cu	Zn	Cd
Armadillidium vulgare	396	397	4.8
Ligia oceanica	96	183	26
Oniscus asellus	103	105	17
Philoscia muscorum	314	111	27
Sphaeroma hookeri	182	124	1.4
Orchestia gammarella	65	159	5.6
Pachygnatha clercki	156	333	23
Pardosa purbeckensis (adults)	338	735	28
Pardosa purbeckensis (juveniles)	109	345	28
Trochosa ruricola (adults)	190	309	21
Trochosa ruricola (juveniles)	156	298	55

In Table 2 concentrations are given for animals caught in another brackish marsh in the port of Antwerp, 'Het Groot Buitenschoor'. Cadmium concentration is highest in the spider.

'Het Paardeschor' is another brackish marsh in that port of Antwerp. Again we see the highest cadmium concentratios in the spider (Table 3).

Saeftinghe is a marsh already in the Netherlands but very close to Antwerp (Table 4). In the two hunting spider species cadmium has a level which is almost as high as in the earthworm.

TABLE 2. Copper, zinc and cadmium concentrations in a crab, a woodlouse, a terrestrial amphipod and a spider in 'Het Groot Buitenschoor'.

Species	Cu	Zn	Cd
Carcinus maenas	67	89	-
Ligia oceanica	223	169	2.5
Orchestia gammarella	52	125	2.1
Tegenaria picta	311	412	10

TABLE 3. Copper, zinc and cadmium concentrations (ppm) in a crab, an isopod, a terrestrial amphipod and a spider caught in 'Het Paardeschor' (port of Antwerp).

Species	Cu	Zn	Cd
Carcinus maenas	83	94	2.8
Sphaeroma hookeri	133	135	1.3
Orchestia gammarella	36	243	2.5
Pardosa purbeckensis	209	532	17

TABLE 4. Copper, zinc and cadmium concentrations in an earthworm, a woodlouse, a terrestrial amphipod and three spider species in Saeftinghe (near port of Antwerp).

Species	Cu	Zn	Cd
Lumbricus spec.	21	292	17
Porcellio scaber	334	260	4.2
Orchestia gammarella	56	114	2.9
Erigone longipalpis	57	287	2.2
Pachygnatha clercki	106	292	8.5
Pardosa purbeckensis	165	414	11

In Table 5 we grouped heavy metal concentrations found in the hunting spider *Pirata piraticus* found in several freshwater tidal marshes and in an unpolluted inland brackish marsh. It can be seen that heavy metal concentrations vary widely.
Tables 6 and 7 give the results for two freshwater tidal marshes. Again it can be seen that cadmium concentrations in the spiders are higher than in the other species. Results for two other freshwater marshes are given in Tables 8 and 9. For both all analysed species are spiders. Looking at the cadmium concentrations it can be concluded that like for other groups we find important interspecific differences in heavy metal accumulation. We have to remark here, however, that the two marshes are quite narrow and there might have been intrusion of individuals from nearby unpolluted habitats.

6. Conclusion

From the above given information it can be concluded that it would be worthwhile to include spiders more often in ecotoxicological field studies and other research projects.

TABLE 5. Copper, zinc and cadmium concentrations (ppm) in Pirata piraticus, a cursorial spider species in twelve freshwater tidal marshes along the Scheldt and in an unpolluted inland brackish marsh.

Site	Cu	Zn	Cd
Bazel-Hoogspanning	116	774	-
Durme-Autostrade	123	530	-
Durme-Tielrode	127	495	-
Kijkverdriet	148	503	15
Moerzeke (12)	165	532	14
Moerzeke (15b-16)	135	509	-
Moerzeke (15N)	184	846	21
Moerzeke (16)	200	735	-
Moerzeke (17)	120	466	-
Tielrode (20)	153	563	13
Tielrode (21)	165	477	7.1
Zele	198	787	163
Molenkreek	83	321	-

TABLE 6. Copper, zinc and cadmium concentrations in a spider and a woodlouse occurring in a freshwater tidal marsh (Berlare).

Species	Cu	Zn	Cd
Clubiona phragmitis	172	709	45
Trachelipes rathkei	308	453	2.8

TABLE 7. Copper, zinc and cadmium concentrations (ppm) in two spider species, a woodlouse and a terrestrial amphipod occurring in a freshwater tidal marsh at Moerzeke (15N).

Species	Cu	Zn	Cd
Clubiona phragmitis	116	494	19
Pirata piraticus	184	846	21
Porcellio scaber	114	239	3.6
Orchestia cavimana	62	250	-

TABLE 8. Copper, zinc and cadmium concentrations in three spider species occurring in a freshwater tidal marsh along the Scheldt (Bazel-Hoogspanning).

Species	Cu	Zn	Cd
Pardosa amentata	184	431	19
Pirat piraticus	166	774	-
Tmeticus affinis	103	307	-

TABLE 9. Copper, zinc and cadmium concentrations in four spider species in a freshwater tidal marsh along the Scheldt (Durme-28).

Species	Cu	Zn	Cd
Centromerus sylvaticus	135	97	-
Clubiona phragmitis	201	579	-
Pardosa amentata	121	458	5.6
Pirata piraticus	123	530	-

7. References

1. Clausen, I.H.S. (1986) The use of spiders (Araneae) as ecological indicators, *Bull. Br, Arachnol. Soc.* **7**, 83-86.
2. Clausen, I.H.S. (1989) On the dynamics od cadmium and lead in Steatoda bipunctata, in J. Bohac and V. Ruzicka (eds.), *Bioindicatores deteriorisationis regionis*, South Bohemian Biological Centre, Ceske Budejovice, pp. 315-318.
3. Hopkin, S.P. and Martin, M.H. (1985) Assimilation of zinc, cadmium, lead, copper and iron by the spider Dysdera crocata, a predator of woodlice, *Bull. Environ. Contam. Toxicol.* **34**, 183-187.
4. Hopkin, S.P. (1989) *Ecophysiology of metals in terrestrial invertebrates*, Elsevier Applied Science, London.
5. Hopkin, S.P. (1993) In situ biological monitoring of pollutants of in ecosystems, in P. Calow (ed.), *Handbook of ecotoxicology, Volume 1*, Blackwell, Oxford, pp. 421-437.
6. Hunter, B.A., Johnson, M.S. and Thompson, D.J. (1987) Ecotoxicology of copper and cadmium in a contaminated grassland ecosystem. II. Invertebrates, *J. Appl. Ecol.* **24**, 587-599.
7. Maelfait, J.-P. and Baert, L. (1975) Contribution to the knowledge of the arachno- and entomofauna of different woodhabitats. Part I. Sampled habitats, theoretical study of the pitfall method, survey of the captured taxa, *Biol. Jb.*

Dodonaea **43**, 179-196.

8. Maelfait, J.-P., Jocqué, R., Baert, L. and Desender, K. (1990) Heathland management and spiders. *Acta Zool. Fennica* **190**, 261-266.
9. Maelfait, J.-P., De Knijf, G., De Becker, P. and Huybrechts, W. (1994) Onderzoek naar de relatie tussen waterhuishouding en spinnenfauna van het Walenbos (Tielt-Winge, België), *Nwsbr. Belg. Arachnol. Ver.* **9(3)**, 59-71.
10. Van Straalen, N.M. and Van Wensem, J. (1986) Heavy metal content of forest litter arthropods as related to body-size and trophic level. Environ. Pollut. **42A**, 209-221.

SOIL ANIMALS AND BIOINDICATION

H. H. KOEHLER
University of Bremen, FB2
Institute of Ecology and Evolutionary Biology
D-28334 Bremen
Germany

Abstract

This chapter focusses on the identification of bioindicators from an ecosystem-oriented perspective. An organism interacts with its biotic and abiotic environment. Consideration of these interactions helps to select organisms suitable for indication. Exposure pathways at the individual level and higher hierarchical levels, up to the ecosystem, are addressed. Five indicator systems are briefly reviewed. From the ecosystem-centred view, the following procedure of risk assessment is proposed: selection of a group of potentially affected organisms, performance of toxicological tests on some members of this specific group, and extrapolation with models. The results of this hierarchical approach and the ecological expertise added to it feed the iterative interplay between science and policy for the establishment of critical concentrations.

1. Introduction

There is a continued search for indicators of stress or more specific impact, because a holistic approach considering all organisms of an area under consideration is unfeasible. Soil animals are used as bioindicators: 1) in risk assessment of known chemicals (prediction), mainly applied to agrochemicals, and 2) in impact assessment of unknown or diffuse stresses, often chronic. It is almost trivial to add that bioindication with soil animals applies only to the animated zone of the earth's crust, an important limitation when effects on groundwater are suspected.

One group of agricultural chemicals causing problems are pesticides, which are used as subsidy or even substitute for natural control (competition, predator-prey interactions) in deliberately altered ecosystems; at the expenditure of energy, agricultural systems are optimized for the crop, not for ecological control processes. In this

N. M. van Straalen and D. A. Krivolutsky (eds.), Bioindicator Systems for Soil Pollution, 179–188.

situation, populations of organisms may become unstable and compete with man for the crop. Control of pest populations using chemicals may harm the truncated agricultural biocoenosis and endanger its remaining control potential, including rate regulation of bioelement cycling.

Models based on NOEC studies with a limited number of organisms may be used to assess the risk of a known xenobiotic for non-target organisms [1]. Another approach relies on a set of selected species (a "guild" in the broadest sense), that indicates changes in the non-target biocoenosis. Criteria for identifying such a guild have to be derived from ecological knowledge and theory [2, 3].

The ecology-centered approach mentioned above also applies to stress assessment of unknown or diffuse impact(s). Bioindicators or indicator systems may also be used for the assessment of chronic environmental problems.

2. Identification of bioindicators in soil

An organism interacts with its biotic and abiotic environment. Consideration of these interactions helps to select organisms suitable for the indication of effects, *e.g.* of classes of chemicals. The indication of impacts of water-soluble toxicants in soil, for example, should focus on

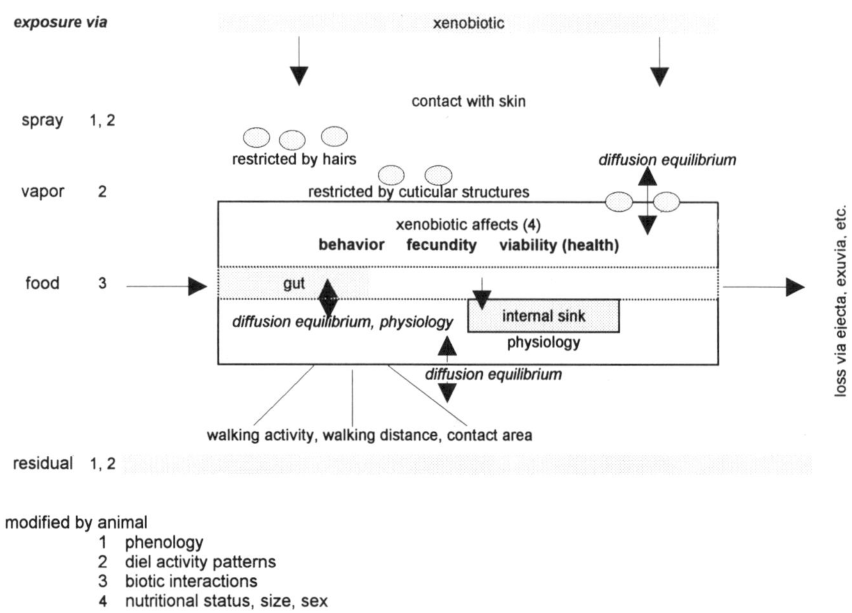

Figure 1. Conceptual model of a soil animal exposed to a xenobiotic chemical.

organisms of the pore-water, or the water film of pores, plus their specific food and their predators. With this ecosystem perspective, extrapolation becomes feasible to higher hierarchical levels (population, guild, community, zoocoenosis, biocoenosis, ecosystem), and to larger scales in space and time. On all levels, compatibility of scales in space and time must be observed with great care. Organisms with specific ecological requirements (stenoecious organisms) have the best indicator qualities for changes of the environment.

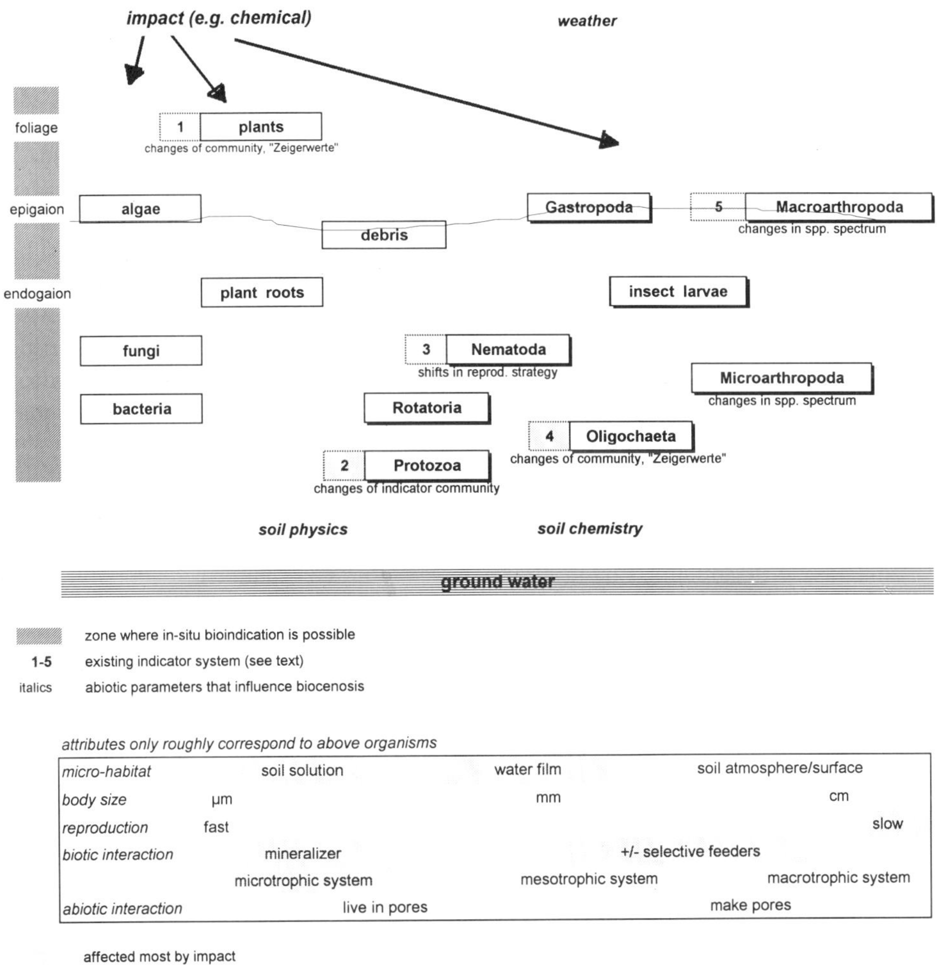

Figure 2. Schematic representation of a soil ecosystem exposed to chemical impact, showing the various biotic components, their depth distribution in the soil profile (above), and the scales of their habitats (below).

A xenobiotic chemical affects an organism according to its specific toxicity, concentration and exposure. For soil animals, the latter depends on mode of application, structure of plant cover, physical and chemical features of the toxicant and the soil, and on morphology and biology of the organism [4]. This is illustrated in Fig. 1. The toxicant may affect life processes negatively (*e.g.* reduced fecundity), or positively (*e.g.* increased fecundity, hormoligosis, see [5, 6]).

Endogeic soil animals are found in three major microhabitats: water of water filled pores, water films of aerated pores, and air-filled lumina of pores (Fig. 2). Contact with water soluble substances may lead to direct intoxication of the first two groups, whereas spray, vapor and uptake with food may affect the last group. For the latter group, exposure pathways may be assessed using a more detailed grouping, for example a classification based nutritional biology, like in food-web systems [7]. Three main trophic groups are: 1) the **microtrophic system** in the water-film, which comprises organisms using easily available C-resources (bacteria, yeasts) and their consumers and predators (Protozoa, Rotatoria, Nematoda), 2) the **mesotrophic system** with users of a wide spectrum of substrates (fungi) and their consumers and predators (mesofauna), and 3) the **macrotrophic system** with animals that are able to dig their own pores and feed on a wide spectrum of microflora and microfauna, detritus plus their predators. In many cases these guilds are, however, not clear-cut, because several animals feed on a variety of resources and there is a lack of knowledge on their nutritional biology.

A wide spectrum of organisms has been used for indication [8] and effects varying from extinction to mass development have been observed after chemical application [9, 10]. Approaches in which a group of organisms with similar responses to chemicals are considered seem to be suitable for individual-oriented simulation.

3. Extrapolation to higher hierarchical levels

Conceptual ecosystem models (like in Fig. 2), and the ecological knowledge contained in them, form the basis for extrapolation of effects observed on indicator species to higher hierarchical levels and for the assessment of consequences for ecosystem persistence.

In ecotoxicological field research [10, 11] ecosystem behaviour after an impact or under chronic stress has been assessed (see Fig. 3). The response of the system is expressed in some parameter that measures ecosystem quality. The definition of quality may depend on the objective of the study and the nature of the system considered. The impact causes a bifurcation between ecosystems exposed to different degrees (Fig. 3). In highly affected systems, successional pathways are altered, and it is a matter of time and scale whether recovery is achieved (scenario 3 in Fig.

3) or not. The establisment of recovery depends also on the research effort applied and the allowance for variability and singularities.

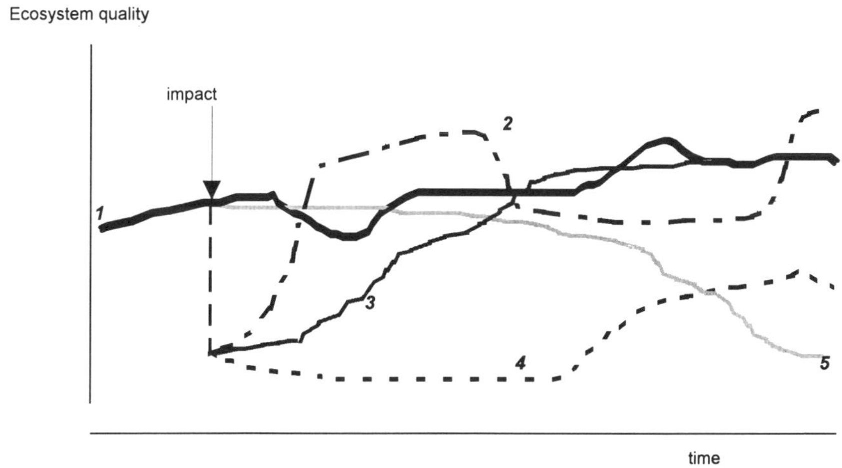

Figure 3. Five possible scenarios of ecosystem development following a disturbance. Scenario 1: control, arbitrarily assumed to increase slightly with time, scenario 2: impact followed by an overshoot and return to control, scenario 3: impact followed by gradual recovery to control, scenario 4: impact followed by a long-term negative deviation from the control, scenario 5: chronic effects, appearing long after the disturbance.

4. Indicator systems

The term indicator system is used here for established bioindication tools based on groups of organisms, for which sufficient ecological knowledge has been collected to relate community data to environmental conditions. Five of such systems are discussed below, with reference to Fig. 2.

4.1. PLANTS

For vascular plants, ecological requirements have been assessed empirically and systematically since the middle of this century to establish a system of "Zeigerwerte". A plant sociological investigation using the "Zeigerwerte" gives complex information on, *e.g.*, soil conditions and heavy metals [12]. Plants are exposed to xenobiotic chemicals on various scales and through a variety of pathways: spatial scales (above and below ground) in the order of of meters, temporal scales in the order of years, exposure via air (shoot), water, and solid residues (root). Rather large areas can be "sampled" by direct observation. Taxonomy is well developed.

4.2. PROTOZOA

Since the beginning of this century, limnologists have developed a sophisticated indicator system mainly using qualitative criteria of the protozoan community (saprobic system) [13]. The organisms are small and their lifetime is short, but they are very numerous and they exist in communities persisting under specific environmental conditions. Exposure to chemical substances is via water. Sampling may cause statistical problems, particularly in soil, and may give distorted results. Qualitative aspects are preferred. Taxonomy starts to develop [14].

4.3. NEMATODA

Bongers [15] developed the maturity index. He proposed to use the r/K-concept to arrange nematodes according to colonizers asd persisters. Communities under stress show a high percentage of colonizers (r-strategists) in contrast to unstressed communities, which are dominated by persisters (K-strategists). The system is not specific for xenobiotics. Considering the possible exposure pathways, water soluble substances may be best indicated. Temporal and spatial scales are rather small. Mixed samples are taken and taxonomy is well developed [16].

4.4. OLIGOCHAETA

Oligochaeta (Lumbricidae and Enchytraeidae) have been found to be good indicators for the decomposer community. A system has been developed by Graefe [17], following closely the plant system. The oligochaete community is predominately affected by humus type. Direct exposure is to water soluble substances. Scales vary widely and reach values similar to those of plants. Sampling may cause statistical problems and qualitative aspects are emphasized (see [5]). Taxonomy for Enchytraeidae is difficult, for Lumbricidae well developed.

4.5. MACROARTHROPODS

Living mainly on the soil surface, macroarthropods are exposed directly to spray and vapor, but also indirectly via their food. Changes in their environment often result in an unspecific reaction. For Staphylinidae (rove beetles), an index to estimate disturbance has been proposed by Bohac [18]. For its calculation, the species in a rove beetle community are sorted according to three levels of increasing anthropophily. Long-term effects may be overcome by the beetles' ability to reinvade disturbed areas from undisturbed areas. An interesting approach of assessing anthropogenic stress is the concept of fluctuating asymmetry, which does not need any control or reference situation [19]. This method may also be used for microarthropods. Time scales cover years and spatial scales have the dimensions of fields or greater. Sampling (*e.g.* pitfall trapping) may be highly biased. Taxonomy is well developed.

4.6. OTHER GROUPS

As animals of the air-filled soil pores, microarthropods are not so much exposed to water soluble xenobiotics, but more to very fine spray and vapor. The main exposure route is indirect, via the food. Karg [20] advocates the use of predatory mites (Gamasina) as indicators, since they integrate complex effects on a lower trophic level. Weigmann and Kratz [21] published some ecological characteristics of Oribatida, on which a bioindicator system may be based. Dunger [22] gives several prerequisites for bioindicators, which are fulfilled by Collembola. Spatial and temporal scales for soil microarthropods range between those for Nematoda and Oligochaeta. Sampling strategies of 10 replicates and clumped distribution patterns of microarthropods lead to semi-quantitative data, but allow good estimates of species spectra [23]. Sampling procedures and statistics have to be improved. Taxonomy is well developed [24].

Molluscs react sensitive to changes in the environment, partly due to their very limited mobility. Many species, however, particularly the long living ones, respond with characteristic time lags. Persistence of shells, a well developed taxonomy and a rather extensive ecological knowledge (listed in 20 categories for 276 species) are a good basis for the development of an indicator system [25].

5. Conclusions

Soil animals live in a very complex environment and have adapted to these conditions very specifically. Testing tools have been developed for laboratory experiments (microcosms), composed of a small number of organisms, selected according to their representativeness for key groups in the ecosystem. Another approach would be to develop indicator systems that use biologically and ecologically homogeneous groups of organisms, *e.g.* selected according to microhabitat, nutritional biology or potential exposure. Not a single, perhaps mythical "most sensitive" indicator species is assessed [26], nor the whole community [1], but a set of populations with relatively clear and similar ecological requirements (a guild in a broader sense), composed of an array of species, sufficiently diverse to allow the detection of changes. Prerequisites for the erection of these indicator systems are taxonomical, biological and ecological knowledge of the guild. Unfortunately, this ecological basis may be rather narrow, which may lead to oversimplification and wrong assumptions [27]. Advanced indication systems with sufficient empirical background to establish standards for various ecosystems may be applied without a reference situation.

The question which approach is most appropriate depends on the problem addressed. The evaluation of the Bioindicator Project funded by the German Ministry of Research and Technology [28] shows that universal indicators do not exist and that reactions of parameters vary considerably, depending, *e.g.*, on the species observed or on the hiearchical level [10, 29]. Mathes and Weidemann [28] also stress the importance of validation of laboratory findings in the real world, based on a sound ecosystem theory.

Guilds indicate an impact or a change of their environment, but an observed effect does not give answers concerning critical concentrations. Indication of an impact through guilds, selected adequately according to considerations of the ecosystem context, has to be combined with risk assessment using models for establishing maximum acceptable concentrations; the models should support extrapolations from toxicological studies conducted only with members of the specific guild. The ecosystem perspective determines the representativeness of the guild, while appropriate NOEC studies (which can be done with members of the guild only), and model extrapolations provide quality criteria for policy. The complexity of this approach, and its intimate link with scientific progress, promises relative safety, provided that an intensive iterative interplay between science and policy is maintained.

Acknowledgements. The presentation of this report was supported by a travel grant of the FNK, University of Bremen.

6. References

1. Van Straalen, N.M. and Denneman, C.A.J. (1989) Ecotoxicological evaluation of soil quality criteria, *Ecotox. Environ. Safety* **18**, 241-251.
2. Jepson, P.C. (1993) Ecological insights into risk analysis: the side effects of pesticides as a case study, *Sci. Total Environ.* **Suppl.**, 1547-1566.
3. Mathes, K. and Weidemann, O. (1990) A baseline-ecosystem approach to the analysis of ecotoxicological effects, *Ecotox. Environ. Safety* **20**, 197-202.
4. Croft, B.A. (1977) Susceptibility surveillance to pesticides among arthropod natural enemies: modes of uptake and basic responses, *J. Plant Diseases and Protection* **84**, 140-157.
5. Luckey, T.D. (1968) Insect hormoligosis, *J. Econ. Entomol.* **61**, 7-12.

6. Neubauer, I.B., Raccah, N., Aharonson, E., Swirski, I., and Ishaaya, I. (1983) Systemic effect of Aldicarb, Dimethoate and Ethiofencarb on mortality and population dynamics of the spirea aphid Aphis citricola Van der Groot, *Crop Protection* **2**, 211-218.
7. Heal, O.W. and Dighton, I. (1985) Resource quality and trophic structure in the soil system, in A.H. Fitter (ed.), *Ecological Interactions in Soil*, Blackwell, Oxford, pp. 339-354.
8. Paoletti, M.G., Favretto, M.R., Stinner, B.R., Purrington, F.F., and Bater, I.E. (1991) Invertebrates as bioindicators of soil use, *Agric. Ecosyst. Environ.* **34**, 341-362.
9. Edwards, C. and Thompson, A.R. (1973) Pesticides and the soil fauna, *Res. Rev.* **45**, 1-79.
10. Koehler, H. (1994) A case study on bioindication and its use for the assessment of ecological impact, in M.H. Donker, H. Eijsackers and F. Heimbach (eds.), *Ecotoxicology of Soil Organisms*, Lewis Publ., Boca Raton, pp. 427-444.
11. Koehler, H. (1992) The use of soil mesofauna for the judgement of a chemical impact on ecosystems, *Agric. Ecosyst. Environ.* **40**, 193-205.
12. Ellenberg, H., Düll, R., Wirth, V., Werner, W., and Paulißen, D. (1991) Zeigerwerte von Pflanzen in Mitteleuropa, *Scripta Geobot.* **18**, Gottingen, 248 pp.
13. Kolkwitz, R. and Marsson, M. (1909) Ökologie der tierischen Saprobien. Beiträge zur Lehre von der biologischen Gewasserbeurteilung, *Int. Revue ges. Hydrobiol. Hydrogr.* **2**, 126-152.
14. Foissner, W. (1993) *Colpoda (Ciliophora)*, Gustav Fischer Verlag, Stuttgart.
15. Bongers, T. (1990) The maturity index: an ecological measure of environmental disturbance based on nematode species composition, *Oecologia* **83**, 14-19.
16. Bongers, T. (1988) *De Nematoden van Nederland*, Koninklijke Nederlandse Natuurhistorische Vereniging, Utrecht.
17. Graefe, U. (1993) Die Gliederung von Zersetzergesellschaften für die standortsökologische Ansprache, *Mittlgn. dtsche. bodenkdl. Ges.* **69**, 95-98.
18. Bohac, J (1990) Numerical estimation of the impact of terrestrial ecosystems by using the staphylinid beetles communities, *Agrochem. Soil Sci.* **39**, 565-568.
19. Palmer, A.R. and Strobeck, C. (1986) Fluctuating asymmetry: measurement, analysis, patterns, *Ann. Rev. Ecol. Syst.* **17**, 391-421.
20. Karg, W. (1968) Bodenbiologische Untersuchungen über die Eignung von Milben, insbesondere von parasitiformen Raubmilben, als Indikatoren, *Pedobiologia* **8**, 30-39.

21. Weigmann, G. and Kratz, W. (1981) Die deutschen Hornmilbenarten und ihre ökologische Characteristik, *Zool. Beitr.* **27**, 459-489.
22. Dunger, W. (1982) Die Tiere als Leitformen für anthropogene Umweltveränderungen, *Decheniana* **26**, 151-157.
23. Ekschmitt, K. (1993) Richtwerte für die Datenqualität von bodenzoologischen Freilanderhebungen, *Verh. Ges. f. Ökol.* **22**, 451-456.
24. Karg, W. (1993) *Acari (Acarina), Milben, Parasitiformes, Cohors Gamasina* Leach, *Raubmilben*, Gustav Fischer Verlag, Jena (2nd. Ed.).
25. Falkner, G. (1990) Vorschläge für eine Neufassung der roten Liste der in Bayern vorkommenden Mollusken, *Schriftenreihe Bayer. landesamt. f. Umweltschutz* **97**, 61-112.
26. Cairns, J. Jr. (1986) The myth of the most sensitive species, *BioScience* **36**, 670-672.
27. Walter, D.E., Kaplan, D.T., and Permar, T.A. (1991) Missing links: a review of methods used to estimate trophic links in soil food webs, *Agric. Ecosyst. Environ.* **34**, 399-408.
28. Mathes, K. and Weidemann, G. (1991) Indikatoren zur Bewertung der Belastbarkeit von Ökosystemen. *Ber. aus der ökol. Forschung* **2**, 126 pp.
29. Koehler, H. (1993) Bodenzoologie (Mesofauna) - Bodenmikrobiologie: Erfahrungen aus interdisziplinärer Arbeit und Möglichkeiten einer Synthese im ökosystemaren Kontext, *Mitt. dtsch. Ges. allg. angew. Ent.* **8**, 555-561.

SOIL FAUNA AS BIOINDICATOR OF RADIOACTIVE POLLUTION

D.A. KRIVOLUTSKY
Institute of Evolutionary Animal Morphology and Ecology
Russian Academy of Sciences
Leninsky Prospekt 33
117071 Moscow
Russian Federation

Abstract

An analysis was made of biodiversity changes and long-term ecological after-effects of radioactive pollution in the former USSR (Russia, Ukraine, Kazachstan) from 1968 to 1994. All types of radioactive pollution have strong negative effects on permanently soil-dwelling invertebrates, mammal populations, and bird reproduction. An initial reduction of animal biodiversity and community structure in Southern Ural (Kyshtym) and in the Chernobyl area was observed, followed by a slow return to the previous conditions. Secondary changes and side-effects on the most active migratory populations of insects, birds and mammals were recorded. The most convenient bioindicators for assessing the ecological effects and environmental quality of sites polluted by radioactive materials appear to be non-migratory animal populations, such as reptiles, earthworms, myriapodes, woodlice, and soil microarthropods.

1. Introduction

For a study on the effects of ionizing radiation on terrestrial ecosystems it is essential to make observations on soil fauna. These comprise 90-95% of all animal species and 90-99% of the total animal biomass. Soil dwellers often prove to be an important factor in forest pest control, especially in radiation-affected ecosystems. Terrestrial carnivorous arthropods are of special importance for the suppression of foci of animal diseases that may develop in areas affected by radionuclide pollution. These carnivores take away many parasites of vertebrates (especially mites) that spend a certain period of their life-cycle in the soil. A close

N. M. van Straalen and D. A. Krivolutsky (eds.), Bioindicator Systems for Soil Pollution, 189–196.

connection between the level of radioactive pollution and the development of agricultural pests is also observed.

Soil fauna have high and stable population densities, amounting to many hundreds of thousands of individuals per m^2 of soil. Since they also have specific feeding habits (phytophageous, saprophageous and canivorous), soil arthropods present a convenient subject for elucidation of the ways and quantitative patterns of radionuclide transfer in communities, and they may be used as bioindicators of polluted territories. The following requirements may be imposed upon animals to be used as bioindicators of industrial pollution by radioactive isotopes [1]: 1) high density and high level of metabolism, 2) long life-span, 3) permanent contact with the anthropogenic factor under study, 4) possibility to collect mass samples, 5) sensitivity to the factor concerned, and 6) fairly large size to facilitate dissection.

Among radionuclides, isotopes that have properties resembling those of the biologically essential elements (^{3}H, ^{14}C, ^{106}Ru, ^{131}I, and ^{137}Cs) are readily incorporated into animal tissues, are easily transferred along food-chains and are of great importance in determining pathways. Fission products of uranium and plutonium, particularly long-lived ^{90}Sr and ^{137}Cs (with radiochemical half-lives of 28.4 y and 30 y, respectively), have always attracted the attention of ecologists. These isotopes are found on the ground surface after nuclear weapon tests, as well as after the processing of the uranium-containing ores and re-processing of nuclear fuel. Many radionuclides, including ^{90}Sr and ^{137}Cs, are rapidly assimilated in detectable amounts into the ecological cycles and can be accumulated by animals. The radiation emitted by these radionuclides is, however, not high in comparison with the natural background and deleterious effects are usually not observed.

Epigeic and soil-dwelling animals associated with the ecological migration of elements, including man-made radionuclides, can serve as one of the biological indicators of environmental pollution with these toxic elements or substances. At the same time, the labeling of ecological cycles in the biosphere with man-made radionuclides opens up new possibilities and prospects for field ecology. This allows us to define pathways of substances in ecosystems more precisely, to clarify trophic links between organisms, and to estimate their productivity.

2. The importance of dosimetry in the assessment of radionuclide effects in the field

Since the ^{90}Sr content in the above-ground parts of a forest stand is much lower than in the soil, radiation doses of surface dwelling invertebrates proved to be much lower than doses received by the soil fauna. Doses in the soil, however, are not sufficiently high to produce a direct damaging effect to invertebrates, at least not to the larval and adult stages. The observed decrease in population density may therefore be attributed to the fact that forest insects (including micro- and mesofauna) hibernate in the soil as eggs. The embryonic development in the egg stage is the most sensitive process in the life-cycle. In this case, irradiation doses are accumulated during several months and may be as great as 200-300 rad.

In laboratory experiments irradiation doses from several hundreds to several thousands rads may cause destruction of eggs and sterility of males and females in some species. Under field conditions, irradiation damage is combined with effects of some other adverse factors (*e.g.* low temperature). The combined effect may cause death at much lower radiation doses compared with doses under favorable conditions.

It should be taken into account that the irradiation dose can be considerably increased (up to 50 %) by internal irradiation, caused by ^{90}Sr incorporated in the tissue. For example, earlier research has documented cytogenetic effects of ^{90}Sr incorporated in *Crepis capillaris* seeds. Internal irradiation has a tenfold more destructive effect compared to external radiation, however, the absolute values of the incorporated irradiation doses in animals are not high; they are 2-3 orders lower than doses accumulated due to external radiation.

Another example illustrating the importance of dosimetry for the interpretation of radionuclide effects is the exposure received by earthworms. These animals are not only irradiated from the outside but also from the soil they swallowed. Most of the other terrestrial animals do not ingest soil and consume only animal or vegetative food in which the radionuclide content is considerably lower (by 1-2 orders of magnitude) than in the soil.

Earthworms collected from a radium polluted soil showed considerable changes in the structure of the integument epithelium (in particular the shape and size of mucous cells). The mucous cells of these earthworms were broader and larger as compared with controls and the number of mucous cells in the radiation-affected earthworms was much higher than in the control ones. There were also less cambial cells in cross sections of earthworm tissue, a large number of regeneration nests in the mid-gut epithelium, and individual regeneration cells in the basal part of the epithelium.

Radioactive pollution exerts the greatest influence on the permanent soil dwellers by decreasing their population level, especially in deep soil layers, and by abrupt disturbances in the structure of animal communities. Temporary soil dwellers are possibly vulnerable to some secondary effects (influence of heat on predators, parasites, fodder plants, *etc.*), rather than to the direct influence of radiation itself. They may not only show a quantitative decrease, but also a higher population density.

Observations on Protozoa in control and ^{90}Sr-polluted plots revealed differences in protozoan numbers and species composition. The effects of radioactive pollution on Protozoa could, however, not be explained by their inherent sensitivity to irradiation (they are actually very resistant), but by the high radiation level in water films around soil particles with absorbed ^{90}Sr. In addition, the differential survival of the species was determined by the presence of dormant life-stages, rather than by differences in the resistance to irradiation.

The examples given above emphasize the importance of conclusive dosimetric evidence, which is necessary for the interpretation of observed effects of irradiation on invertebrate populations in polluted ecosystems. More information on radiation-induced mortality in different species of soil animals is required to estimate the ranges of sensitivity of species not examined before. Such information would allow the assessment of environmental problems after uncontrolled or deliberate emissions of radionuclides in the environment (calamities, nuclear attack); it may also be useful for investigations on the control of invertebrates by irradiation.

3. Food-chain transfer of radionuclides

According to some radioecologists, it is necessary to elucidate the peculiarities of the transfer of an element from one trophic level to another, including element accumulation factors in organisms and the degree of retention of elements by organisms. In spite of the abundant radioecological literature on the migration of elements in trophic chains, the mechanisms underlying the transfer processes have, however, not been studied enough to build a suitable universal model [2]. In particular, the coupling between elements during migration in trophic chains is one of the most important and poorly studied factors [3]. In experiments with laboratory and domestic animals, the problem of non-radioactive isotope effects on radionuclide accumulation are considered in detail, but for natural populations and ecosystems this problem is not solved, even not for well-studied pairs of elements such as Ca - ^{90}Sr and K - ^{137}Cs.

Field studies on the co-migration of elements in food chains allow us to identify some fundamental aspects of element concentrations in animal bodies [4]. These may be summarized as follows: the concentration of macroelements in animal bodies is not related to the taxonomic classification system, with the exception of Ca and S. With respect to calcium, there are groups of animals that concentrate much more Ca than others: molluscs, diplopods, woodlice, and vertebrates to some extent.

Microelement concentrations (including Sr and Cs) in animal bodies are determined by species-specific factors on the one hand, and by the concentration in the food on the other hand. The higher the microelement concentration in food, the higher their concentration in animal bodies. The dependency of the element content in bodies upon the element content in food generally increases with the atomic number of the element.

The range of element concentrations in organisms is determined by the physiological function of the element and the possibilities for organisms to tolerate a given element concentration. Not all chemical elements found in organisms have, however, a particular biochemical function. Some of them are just turning over as "ecological impurities", usually associated with carriers. As long as the carrier is loaded with a low amount of the "ecological impurity", there will be no damage, but at a certain level of substitution toxic effects may become manifest.

The structure of a trophic chain also has an important effect on the transfer of elements through it. Recent studies of element turnover and their migration through animal populations allowed the proposal of a new conceptual model for a trophic chain, which includes three heterotrophic levels: substrate - microorganisms - phytosubstrate [animals (phytoconsumers)] - animal substrate [animals (zooconsumers)]. A bifurcation of flows occurs for the alkalines, the alkaline-earth elements and microelements in saprophageous and herbivorous animals. In this case, one fraction of the element is assimilated in animal bodies from the substrate. The other penetrates through populations of microorganisms of the digestive tract. The important practical aspect of bifurcation of flows in the gastro-intestinal tract is that with increasing flow through the microorganism community the accumulation in animal bodies also increases.

With respect to the features of element concentrations in animal bodies and food chains let us return to the transfer of element pairs in food chains. It is obvious that Sr behaves relative to Ca as an "ecological impurity" and its relative concentration decreases as it migrates in a trophic chain. As for Cs, it is not as closely connected with K as Ca is with Sr. In vertebrates with carnivorous or omnivorous food patterns, Cs transfer from diet to organisms is more affected by digestive features than by its relationship with the carrier element.

The distribution of Sr and Cs in the biomass of particular components of an experimental ecosystem corresponds to the ratio of element pairs at the different trophic levels, but there are some deviations [1, 4]. Microarthropods accumulate Cs and Sr more intensively and earthworms less strongly than would be expected from the distribution of K and Ca in their biomass, and from the contents in substrates.

4. Ecological effects of radioactive pollution at Chernobyl and Kyshtym

Extensive field studies were carried out during 1986 to 1993 by standard soil-zoology quantitative methods in the framework of an ecological research programme on the short-term and long-term after-effects of the accident with the Chernobyl Atomic Power Station in 1986. Observations were done within the protection area and in plots serving as a relative control (70 km south of the station).

In July 1986 samples were collected from soil and litter of pine forest stands, 50-60 years old, with a forest floor consisting of a thick needle litter, a moss layer and a humus layer of 1-2 cm depth. The plots were 3 km away from the station (Izumrudnoe) and 30 km south of the station, near the village of Lutezh. Animals from core samples taken at various depths were extracted by simple Tullgren funnels.

The effects of a high level of natural radioactivity from ^{226}Ra, ^{232}Th, and ^{238}U on soil invertebrate populations were investigated in plots with a a different origin of radioactivity. During a sampling campaign covering 15 seasons, invertebrate populations were monitored on the basis of catches obtained by hand-sorting and thermo-eclectors. The ^{226}Ra concentration in soil varied from 56 to 710 pg/g. Control plots were chosen with similar soil vegetation and a natural background radiation.

Some polluted plots were on a terrace with meadow vegetation, along a river bank. High radiation levels in the soil of these plots were due to the flooding of contaminated groundwater over an area of 15-20 km length and 3-4 km width. The concentration of ^{226}Ra in the soil plots varied from 7.1 to 5.6 pg/g.

TABLE 1. Oribatid mites extracted from soil cores collected in pine forests at different distances from the Chernobyl nuclear power station, in July 1986. Densities (means ± standard errors) are per 225 cm^2

Species and groups	Distance		
	3km	30 km	70 km
Litter dwellers			
No. species in 10 cores	5	22	23
Total density	2.2	140	104
No. species found in all cores	none	5	6
Soil dwellers			
Opiella nova	1.0 ± 0.4	73 ± 23	42 ± 7
Quadrioppia quadricarinata	1.3 ± 1.0	15 ± 8	23 ± 5
Suctobelba sp.	3.6 ± 1.9	52 ± 23	63 ± 20
Tectocepheus velatus	1.4 ± 0.6	27 ± 4	56 ± 11
No. species found in all cores	none	4	4

In the polluted zone at Chernobyl, die-back of pine forests (*Pinus silvestris*) exposed to a high radiation level (absorbed dose varying from 8 to 10 krad) was visually obvious by late summer 1986; the area of dead forest reached about 400 ha. Lower doses (0.8-1 krad) killed part of the trees, mainly those with an age of 10-12 years. Litter invertebrate populations were badly affected by radioactive pollution at a surface dose rate of 7-15 mrad/h. Abundance of soil mites and the early life-stages of all soil fauna were reduced about 30 times by mid-July, 1986. A considerably smaller (2-3 fold) reduction was observed in the populations of soil dwellers in the fields and in the deep layers of the agricultural plots. Presumably, the process of recovery of soil acarofauna began as early as autumn 1986. Larvae and nymphs of soil mites appeared in irradiated soils by October 1986. By autumn 1987 all the main dynamic parameters of the soil acarofauna were comparable to the control [5, 6]. Adult oribatid mites and their juvenile stages dominated soil fauna in that period. Juvenile stages proved to be more sensitive to radioactive contamination than adult animals. Differences between heavily contaminated plots and controls were more clear for the number of juvenile oribatids compared to the adult populations. For mesofauna populations there were no differences between the control plot at 70 km and the plot at the edge of 30 km zone.

In Table 1 data on oribatid mite populations are given for litter and soil dwellers in the control plot and the most strongly polluted sites. Although differences in species composition are found for both primitive and higher oribatid mites, radioactive contamination led to a decrease of population size for both litter and soil dwellers; litter dwellers were influenced by contamination significantly more than soil species.

Another long-term programme on the ecological effects of radioactive pollution was conducted over the years 1969 to 1991, following the accident at Kyshtym, in Southern Ural, in 1957. Observations were done on plots polluted with ^{90}Sr, ^{137}Cs, ^{144}Ce, ^{106}Ru, ^{65}Zn, ^{125}Sb, ^{237}Am, and ^{239}Np.

TABLE 2. Oribatid mite populations in birch forest control plots and in plots polluted by ^{90}Sr in Kyshtym (Southern Ural)

Group	Control 1	Control 2	^{90}Sr
Soil surface and litter dwellers	20	19	5
Mineral soil inhabitants	10	13	9
Total no. species	30	32	14

In the polluted birch forest at Kyshtym, the structure of the oribatid community became simplified due to a loss of many species, see Table 2. The diversity of the mite community markedly decreased, especially with respect to soil surface and litter dwellers. In agricultural plots, a negligible decrease in species composition was noted (9-14 species versus 15 in the control), but there was a two-fold decrease in the population density of adult mites. The difference between the forest and the agricultural plots may be due to a difference in depth distribution of the oribatids. In agricultural soils oribatids are rather evenly distributed over the soil profile, and not so much concentrated at the surface, as in a forest floor.

5. Conclusions

The structure and the dynamics of soil mite communities are directly influenced by soil characteristics. Oribatei can be used as bioindicators of specific soil types, as bioindicators of industrial pollution and as bioindicators of various types of radioactive pollution.

6. References

1. Krivolutsky, D.A. (1983) *Radioecology of Terrestrial Animal Communities*, Energoatomizdat, Moscow (in Russian).
2. Krivolutsky, D.A. and Pokarzhevskii, A.D. (1975) The role of pedobionts in the biogeochemical cycles of calcium and strontium-90 in the ecosystem, in *Progress in Soil Zoology*, Academia, Moscow, pp. 249-254.
3. Pokarzevskii, A.D. and Van Straalen, N.M. (1996) A multi-element view on heavy metal biomagnification, *Appl. Soil. Ecol.* **3**, 95-98.
4. Pokarzevskii, A.D. (1985) *Geochemical Ecology of Terrestrial Animals*, Nauka, Moscow (in Russian).
5. Krivolutsky, D.A. and Pokarzhevskii, A.D. (1992) Effects of radioactive fall-out on soil animal populations in the 30 km zone of the Chernobyl Atomic Power Station, *Sci. Total Environ.* **112**, 69-77.
6. Krivolutzky, D.A., Pokarzhevskii, A.D., and Viktorov, A.G. (1992) Earthworm populations in soils contaminated by the Chernobyl atomic power station accident, 1986-1988, *Soil Biol. Biochem.* **24**, 1729-1731.

A HIERARCHICAL APPROACH TO ECOLOGICAL ASSESSMENT OF CONTAMINATED SOILS AT ABERDEEN PROVING GROUND, USA

R.G. KUPERMAN
Argonne National Laboratory, Environmental Assessment Division
9700 South Cass Avenue, Argonne, IL 60439, USA

1. Abstract

Despite the expansion of environmental toxicology studies over the past decade, soil ecosystems have largely been ignored in ecotoxicological studies in the United States. The objective of this project was to develop and test the efficacy of a comprehensive methodology for assessing ecological impacts of soil contamination. A hierarchical approach that integrates biotic parameters and ecosystem processes was used to give insight into the mechanisms that lead to alterations in the structure and function of soil ecosystems in contaminated areas. This approach involved (1) a thorough survey of the soil biota to determine community structure, (2) laboratory and field tests on critical ecosystem processes, (3) toxicity trials, and (4) the use of spatial analyses to provide input to the decision-making process. This methodology appears to offer an efficient and potentially cost-saving tool for remedial investigations of contaminated sites.

2. Introduction

The field of environmental toxicology has expanded in the past decade; however, most progress has been in studies of aquatic and aboveground terrestrial ecosystems. With few exceptions, soil ecosystems have been ignored in ecotoxicological studies, and in many conceptual models they are presented as a "black box." This may be due to the immense complexity of soil ecosystems and the interdependence of the diverse soil community and the processes it regulates in soil. Methods of studying soil ecosystems have been developed [4] and used by a relatively small group of soil ecologists and are not readily available to environmental toxicologists. Because soil is a net sink for many pollutants and the functioning of the overall terrestrial ecosystem is dependent on soil processes (organic matter decomposition and release of nutrients), it is clear that a comprehensive methodology is needed to enable researchers to accurately assess damage to the soil ecosystem, while providing practical information to decision makers to direct site cleanup and remedial efforts.

N.M. van Straalen and D.A. Krivolutsky (eds.), Bioindicator Systems for Soil Pollution, 197–212.

The proposed methodology is based on a hierarchical approach to assessing the ecological condition of contaminated soils. The methodology incorporates population, community, and ecosystem parameters. The focus of this methodology is on determining changes in the abundance and diversity of soil biota and changes in the rates of critical ecosystem processes that may be affected by soil contamination.

A four-step approach was used to implement a comprehensive assessment program:

1. A thorough survey of soil biota to determine community structure and whether contaminated media are toxic to area biota.
2. Laboratory and field tests on critical ecosystem processes to identify biologically mediated processes in soil most at risk from site contamination.
3. Toxicity trials to determine the relationships between the observed ecological effects in the field and concentrations of soil contaminants.
4. The use of spatial analyses for estimating the extent of the contaminated media with significant adverse ecological effects to provide input to the development of remedial alternatives and cleanup goals.

3. Methods

Soil biota and biologically mediated soil processes were characterized for contaminated soil at J-Field on Aberdeen Proving Ground in Maryland, USA. The objectives were to survey the soil invertebrate community, to conduct an *in situ* litter decomposition assay, and to determine microbial biomass and activity in soils. A summary of assessment and measurement endpoints used for the ecological risk assessment of soil ecosystems is shown in Table 1.

Studies were conducted in the spring, summer, and fall of 1994 in the terrestrial survey grids located in the pushout area near the Toxic Burning Pits (TBTF on Figure 1) and in the local background area approximately 30 m away from the Toxic Burning Pits (TBTC on Figure 1). A reference site with soil parameters similar to those at J-Field was selected in Gunpowder Falls State Park, approximately 7 km west of J-Field. The dimensions of each survey grid were approximately 10 m × 10 m. Samples were collected from 10 randomly selected quadrants (1 m × 1 m) in each of the survey grids. The spatial distribution of soil parameters was determined by sampling along the sitewide survey grid over the entire area of concern.

3.1. SURVEY OF SOIL BIOTA

Determining the abundance and diversity of soil organisms in a given soil ecosystem is a prerequisite to a comprehensive assessment program. Soil invertebrates and microorganisms are essential components of the decomposer food web, nutrient cycling, and environmental

TABLE 1. Summary of assessment and measurement endpoints for ecological risk assessment of soil ecosystems

Assessment Endpoint	Measurement Endpoint	Ecological Component
Community and population parameters	- Taxonomic richness - Abundance/activity	- Soil invertebrate community including macroinvertebrates (Lumbricidae, Gastropoda, Myriapoda, Aranea, Insecta), microinvertebrates (Acari, Collembola, Nematoda), and soil Protozoa - Plant community composition - Aboveground plant biomass
	Abundance of functional groups (trophic structure)	Decomposers, omnivore-predators, bacterivores, fungivores, and herbivores
	Fungal and bacterial abundance and activity	Soil microorganisms, biomass N, CO_2 evolution (SIR)
Effects on critical biological processes	Rates of organic matter decomposition and nutrient release	Litter of *Phragmites australis*
	Nitrogen mineralization	Field incubations
	Enzyme activity	C-, N-, and P-acquiring enzymes

SIR = substrate-induced respiration.

contaminant pathways. They are important agents in the formation and maintenance of the biological, chemical, and physical character of soil ecosystems. Soil biota play a key role in regulating the rates of organic matter degradation and the subsequent release of nutrients.

3.1.1. *Soil Macroinvertebrates*

Three soil cores (9.5-cm diameter and 10-cm depth) were collected with a stainless steel soil sampler in each sampling quadrant on each sampling date. The intact soil was placed in a sorting tray. Animals were extracted from samples by hand sorting and were preserved in ethyl alcohol for further identification. Abundance of soil macroinvertebrates was expressed as the number per square meter (to a depth of 10 cm). Statistical analyses included analysis of variance (ANOVA) on the log-transformed data. The Bonferroni *t*-test was used to determine significance at the $\alpha = 0.05$ level between pairs of test groups.

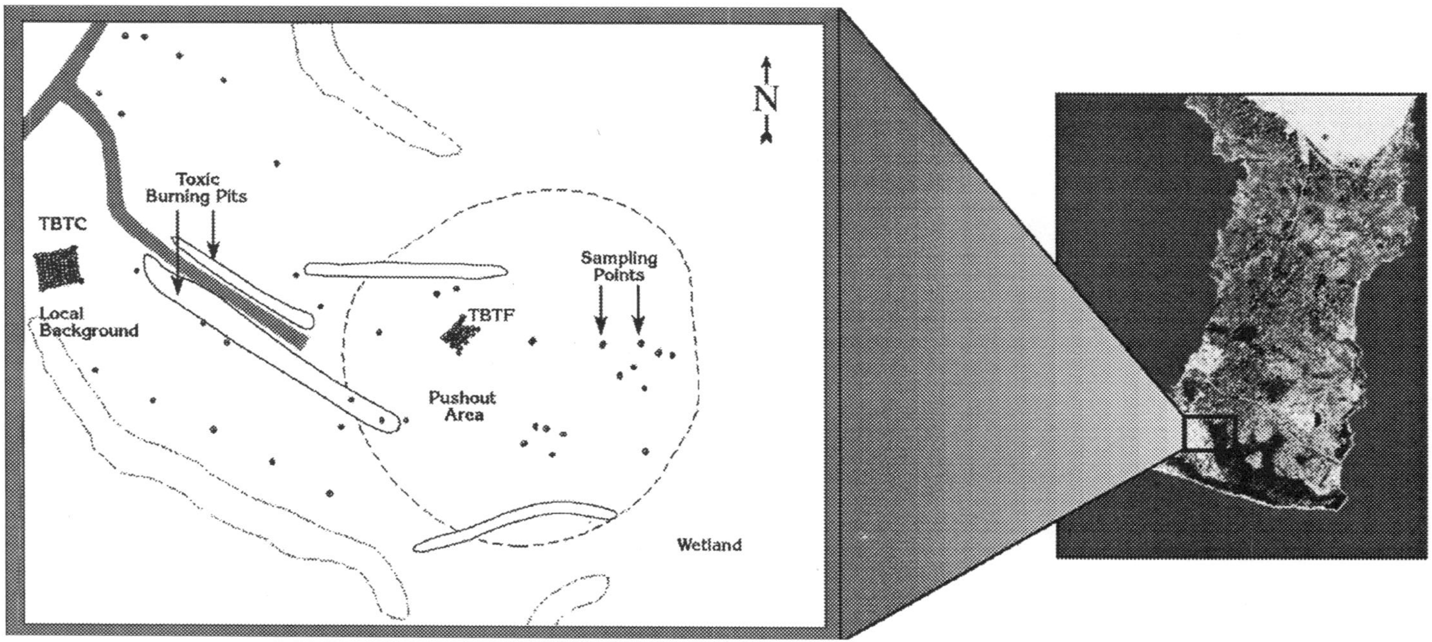

Figure 1. Map of the study area at Gunpowder Neck Peninsula

3.1.2. *Soil Nematodes*

The Baerman funnel technique was used for extraction of nematodes from the soil. This method involves placing a soil sample on a mesh sieve at the wide part of the funnel, which has a rubber tube closed by a clip and filled with water. The nematodes move out of the soil and collect in the neck of the tube, where they can be moved into a dish for counting.

Soil cores (5-cm diameter) were taken to a depth of 10 cm, placed in polyethylene bags, and returned to the laboratory. Twenty-gram subsamples of soil (field moist) were then placed on the extractor (Baerman funnels) for 48 h at room temperature. Nematodes were counted live at 140× magnification and sorted into fungivore, bacterivore, herbivore, and omnivore-predator trophic groups, on the basis of the reported feeding habits and on esophageal morphology.

3.1.3. *Direct Estimates of Active and Total Fungi and Active Bacteria*

Soil microbial biomass was estimated because of the importance of soil microorganisms in nutrient cycling and their role as a source and sink of plant nutrients at J-Field. Microbial biomass is the main acting agent for most soil biogeochemical processes. It interacts with the primary productivity of ecosystems by regulating nutrient availability and degradation pathways of soil contaminants.

For each soil sample, the length of active hyphae stained with fluorescein diacetate (FDA) was measured using epifluorescent microscopy at 250× total magnification [7]. Three separate measurements were taken from the agar film on the slide under the coverslip. Total fungal biomass was measured using the differential interference or phase-contrast microscopy at 250× total magnification. Direct estimates of active bacteria were determined using the 40× objective lens and epifluorescent microscopy by counting the number of FDA-stained (i.e., active) bacteria in each of five individual fields on each slide. Statistical analyses included ANOVA. The Bonferroni *t*-test was used to determine significance at the $\alpha = 0.05$ level between pairs of test groups.

3.2. SURVEY OF BIOLOGICALLY MEDIATED PROCESSES IN SOIL

An assessment of how increased concentrations of soil contaminants may alter rates of litter decomposition as well as the rates and timing of nutrient retention and release is critical to understanding the impact of soil pollution on the overall structure and function of J-Field terrestrial ecosystems. Litter decomposition and nutrient mineralization are important biologically mediated processes affecting soil fertility and primary productivity of ecosystems at J-Field. The distribution and turnover of organic matter contributes to ecosystem functioning by forming soil organic matter pools and nutrient exchange sites for root uptake . Soil enzyme activities and soil respiration were also measured to determine whether soil concentration of contaminants may have detrimental effects on the microbial activity in the study area.

3.2.1. *Litter Decomposition and Nutrient Dynamics in Decomposing Litter*

Decay rates and nutrient mineralization rates of plant residues were determined in the Toxic Burning Pits area using a litterbag method. In this approach, preweighed plant residues are confined in mesh litterbags and placed in the field. Subsets of litterbags are periodically harvested to determine mass loss and change in nutrient content.

Litterbags had a 10-cm × 10-cm inside area and were constructed of fiberglass window screen material with a mesh size of approximately 1.5 mm. Bags were constructed by "welding" together the sides of 12-cm × 22-cm pieces of screen (folded in half) using a flat-tipped soldering iron. Leaf litter of *Phragmites australis* in the litterbags was collected at the time of senescence in October 1993. Plant material was oven-dried at 80°C. Litterbags were filled with 4–5 g of dried plant material and the exact weight recorded. The tops of filled litterbags were sealed and an aluminum tag with an identification number wired to the bag. Subsamples of dried litter were archived for initial litter quality and nutrient analyses.

Litterbags were buried in soil in an appropriate location in the terrestrial survey grids. The locations of litterbags were marked with flags. Extra litterbags were taken to the field, handled as other litterbags, but immediately returned to the laboratory to determine any loss due to handling.

Ten litterbags from each terrestrial grid were collected every six months. Litterbags were placed into labeled "zip-lock" bags as they were collected. Litter was analyzed for total nitrogen (N) and carbon (C) content using the Carlo/Erba NA1500 automated C/N analyzer.

Because of contamination of plant residues with soil, residue masses were corrected for soil infiltration using the following soil correction equation [1]:

$$FLiC = (SaC - SlC)/(LiC - SlC), \qquad (1)$$

where *FLi*C is the fraction of litterbag content that is actually litter C; *Sa*C is the percent of C of the entire litterbag sample; *Sl*C is the mean percent of C of the soil from the field site; and *Li*C is the percent of C of the initial litter residue.

3.2.2. *Soil Enzyme Activity*

This study quantified the activity in soils of beta-glucosidase, carboxymethylcellulase (carbon-acquiring enzymes), n-acetylglucosaminidase (N-acquiring enzyme), and both acid and alkaline phosphatases (phosphorus [P]-acquiring enzymes).

Soil cores were collected with a 5-cm-diameter soil corer to a depth of 10 cm. Assays were run at 25°C on soil slurries suspended in pH 5 acetate buffer. The only exceptions were the samples assayed for alkaline phosphatase activity. These slurries were suspended in buffer at pH 9.5. All the enzymes (except for cellulase activity) were assayed colorimetrically using pNP-linked substrates (pNP = p-nitrophenyl). These substrates were pNP-beta-D-glucopyranoside, pNP-n-acetylglucosaminide, and pNP-phosphate. The cellulase assay was a viscometric assay that used carboxymethyl cellulose as substrate. Statistical analyses included ANOVA on the log-transformed data. A Fisher PLSD was used to determine significance at the $\alpha = 0.05$ level between pairs of test groups ($n = 5$ for each group).

3.2.3. *Soil Microbial Activity*

The soil respiration study measured carbon dioxide (CO_2) evolution in microcosms as an integrator of the effects of soil contamination on the combined activities of microorganisms within the soil ecosystem. A substrate-induced respiration (SIR) method was used to estimate microbial activity in soil. The addition of substrate (glucose) to a soil sample induces a maximal respiratory response from the soil microbial biomass, measured as CO_2 evolution.

Substrate-induced respiration was measured using a soil-respiration measuring system with continuous gas flow [2,3]. The soil-respiration measuring system used an incubation chamber, an air-flow-controlling unit, an air-flow-measuring unit, and a CO_2 analyzer (LI-6251). A 15-g fresh homogenized subsample was placed in a 125-mL Erlenmeyer flask. Glucose solution (60 g L^{-1}) was added to the flask using a syringe to bring the soil water content near to its holding capacity (no free-standing solution). Soil was incubated until the rate of CO_2 evolution from the soil sample become constant at an air flow rate of 180 mL min^{-1} (approximately 40 min). At this point, the CO_2 evolution rate was recorded as the substrate-induced respiration of that soil sample. Statistical analyses included ANOVA. The Bonferroni *t*-test was used to determine significance at the $\alpha = 0.05$ level between pairs of test groups.

3.3. TOXICITY TRIALS

3.3.1. *Earthworm Toxicity Tests*

Earthworm toxicity screening tests utilized the earthworm *Eisenia foetida* in the screening and definitive tests. Survival rates and differences between initial and final weights (sublethal endpoints) were used as indices of toxicity. The test methods used for earthworm toxicity studies were adapted from Karnak and Hamelink [8] and Neuhauser et al. [10]. Analysis of covariance (ANCOVA) was used to test the weight differences, and the *t*-test was used for pairwise comparison of least square means (means adjusted for the covariate) [12]. The screening procedures for determining earthworm toxicity are summarized by Phillips et al. [11].

3.3.2. *Phytotoxicity Tests*

The screening procedures for determining phytotoxicity were adapted from the U.S. Environmental Protection Agency's Early Seedling Growth Toxicity Test [14]. The plant species used in the screening test was lettuce (*Lactuca sativa* [var. Black Seeded Simpson; Meyer Seed Co., Baltimore, MD; lot B1-394]). Data were produced on seed emergence rates, plant heights, survival rates, and plant dry weights. Statistical evaluations of plant data included ANOVA and the Newman-Keuls pairwise comparison of means [12]. The screening methods for all of the plant tests are summarized by Phillips et al. [11].

3.4. SPATIAL ANALYSES

Spatial distributions were predicted using ordinary kriging routines. For all data, a variogram with a range of 27.5 m was fit to the data and the kriging routine okb2d from the

Geostatistical Software Library [5] was used to develop a gridded data set to represent the spatial distribution of the variables. Data values were predicted on a 3-m grid spacing for each variable. Visualization of these data were created using IRIS Explorer [13]. Surfaces were developed using the northing and easting of the predicted grid, the parameter of interest for elevation, and the sum of total lead (Pb), copper (Cu), and zinc (Zn) concentration for shades of grey. The images were rotated to allow the topology to be visible.

4. Results

All field and laboratory procedures are ongoing and cannot be fully presented in this paper. Some results from the study are reported here to illustrate the utility of the comprehensive soil assessment methodology and the advantages of a hierarchical approach for determining the ecological effects of soil contamination.

The mean total concentrations (±S.E.) of seven heavy metals (arsenic, cadmium, chromium, copper, nickel, lead, and zinc) at the reference, local background, and contaminated (pushout) sites were, respectively, 1.75±0.09, 7.7±0.3, and 39.5±3.2 mM kg^{-1} dry mass of soil. Therefore, total heavy metal concentration at the local background site was 4.4 times greater than at the reference site. The contaminated site contained a total heavy metal concentration 22.5 times greater than at the reference site.

In the spring of 1994, the total abundance of macroinvertebrates was significantly ($P = 0.0001$) lower in the pushout area (TBTF) than at the local background site (TBTC). In the fall of 1994, the total abundance of macroinvertebrates did not differ significantly between the two sites, but remained significantly ($P = 0.0001$) lower than at the reference site.

Nematodes were strong indicators of soil contamination in the study area. The numbers of most trophic groups and total nematodes were significantly lower on all three dates in 1994 in the pushout area than at the local background site (Table 2). In April, all trophic groups were affected, and the total numbers of nematodes were reduced by 85% in the pushout area (TBTF) compared with the local background site (TBTC). The toxic effects on the nematode community were not as pronounced in June, but fungivore, herbivore, and hatchling groups were impacted negatively. Fungivore nematodes appeared to be particularly sensitive. The total numbers of nematodes were reduced significantly by 37% in the pushout area compared with the local background site. In October, nematodes were again significantly reduced in the pushout area compared with the local background area. All trophic groups were reduced except the herbivores. Total numbers were lower at the pushout area than at the local background and reference sites.

Microbial biomass was depressed in the local background and contaminated sites. Total fungal biomasses for the local background and contaminated sites were 54% and 15% that of the reference site, and active bacterial biomasses were 71% and 19% that of the reference site (Table 3). Microbial activity as measured by the substrate-induced respiration method was also reduced in polluted soils and paralleled the decrease in the microbial biomass (Table 3).

TABLE 2. Changes in the trophic structure of soil nematodes in areas with high (pushout) and low (local background) concentrations of heavy metals

		Number per Gram Dry Soil		
Season	Trophic Group	Local Background (TBTC)	Pushout Area (TBTF)	*P*
Spring	Fungivore	5.30	0.29	0.0001
	Bacterivore	6.42	1.82	0.0012
	Herbivore	2.49	0.10	0.0004
	Omnivore	0.21	0	0.0160
	Hatchling	0.78	0.05	0.0031
	Total	15.72	2.41	0.0001
Summer	Fungivore	5.11	0.04	0.0001
	Bacterivore	12.99	12.98	0.3812
	Herbivore	3.34	2.08	0.0565
	Omnivore	0.21	0.04	0.2302
	Hatchling	1.59	0	0.0001
	Total	24.02	15.20	0.0567
Fall	Fungivore	3.43	1.09	0.0099
	Bacterivore	11.22	6.47	0.0493
	Herbivore	2.36	4.16	0.8414
	Omnivore	0.31	0	0.0393
	Hatchling	0.85	0.19	0.0166
	Total	18.60	12.06	0.0837

The activity of C-, N-, and P-acquiring enzymes was significantly lower in the local background and contaminated sites than at the reference site (Table 4). The local background and contaminated sites exhibited enzyme activities that were only 14%–25%, and 2%–11% those of the reference site, respectively. All enzyme activities and total fungal biomasses were significantly ($P < 0.001$) correlated for all samples. This provided greater confidence in the reliability of both enzyme and microbial data sets.

TABLE 3. Microbial characteristics of soils in areas with high (pushout) and low (local background) concentrations of heavy metals

Soil Microbial Parameter	Reference Site	Local Background (TBTC)	Pushout Area (TBTF)
Active bacterial biomass ($\mu g\ g^{-1}$ DW)	15.37[a]	10.94[ab]	2.89[b]
Total fungal biomass ($\mu g\ g^{-1}$ DW)	335.27[a]	182.95[a]	48.96[b]
SIR ($\mu g\ CO_2\ min^{-1} g^{-1}$ soil DW)	8.11[a]	2.57[b]	1.83[b]

Note: Numbers with the same letter are not significantly different at $\alpha = 0.05$.

DW = dry weight.

TABLE 4. Soil enzyme activity in areas with high (pushout) and low (local background) concentrations of heavy metals

Soil Enzyme	Reference Site	Local Background (TBTC)	Pushout Area (TBTF)
Beta-glucosidase ($\mu M\ g^{-1}$ DW h^{-1})	0.63[a]	0.10[b]	0.01[c]
Carboxymethyl cellulose (viscometric units g^{-1} DW h^{-1})	87.40[a]	16.30[b]	8.40[b]
N-acetylglucosaminidase ($\mu M\ g^{-1}$ DW h^{-1})	0.27[a]	0.07[b]	0.02[c]
Acid phosphatase ($\mu M\ g^{-1}$ DW h^{-1})	1.78[a]	0.40[b]	0.08[c]
Alkaline phosphatase ($\mu M\ g^{-1}$ DW h^{-1})	0.30[a]	0.12[b]	0.15[b]
Total phosphatase ($\mu M\ g^{-1}$ DW h^{-1})	2.08[a]	0.52[b]	0.23[c]

Note: Numbers with the same letter are not significantly different at $\alpha = 0.05$.

DW = dry weight.

Litter decomposition was significantly faster at the contaminated site (pushout area) than at the background site at both 6 and 12 months (Table 5). Decay was particularly rapid at the contaminated site, and only 6% of the initial C remained after one year. Nitrogen dynamics of decomposing litter exhibited contrasting patterns at the two sites. At the background site, there was significantly more net N mineralization from litter than at the contaminated site after six months. However, by 12 months, whereas the litter at the local background site had begun to show net immobilization of N, litter at the contaminated site had a net mineralization of N. These results strongly suggest that at the contaminated site, litter C and N dynamics were altered, possibly through the alteration of soil biotic activity. The more rapid decay and greater net N mineralization in litter at the contaminated site have yet to be explained.

The results of toxicity trials were reported by Phillips *et al.* [11]. A summary of the results is shown in Figure 2. The outcomes of the earthworm toxicity and the phytotoxicity tests were conflicting. Most of the earthworm toxicity tests indicated that soil was not toxic (except for two locations) in the study area. On the contrary, the phytotoxicity test showed toxic effects in all sampling locations.

The spatial distribution of the sum of Pb, Cu, and Zn concentrations is shown in Figure 3. An example of the spatial correlation of total enzyme activity with the sum of the concentrations of Pb, Cu, and Zn is given in Figure 4. Geostatistical techniques were used to determine the spatial distributions of these parameters and to evaluate various cross-correlations. Geostatistical techniques allow the spatial correlation to be determined for two variables even when sample points were not collocated. The parameter distributions were

TABLE 5. Carbon and nitrogen dynamics in decomposing litter of *Phragmites australis* in areas with high (pushout) and low (local background) concentrations of heavy metals

Time (months)	Parameter	Local Background (TBTC)	Pushout Area (TBTF)	*P*
6	% C remaining	39.82	26.37	0.0660
12		25.22	6.18	0.0001
6	Litter N content (%)	0.81	1.04	0.0343
12		1.29	1.60	0.0622
6	% N remaining	77.85	94.09	0.0789
12		108.84	73.01	0.0005

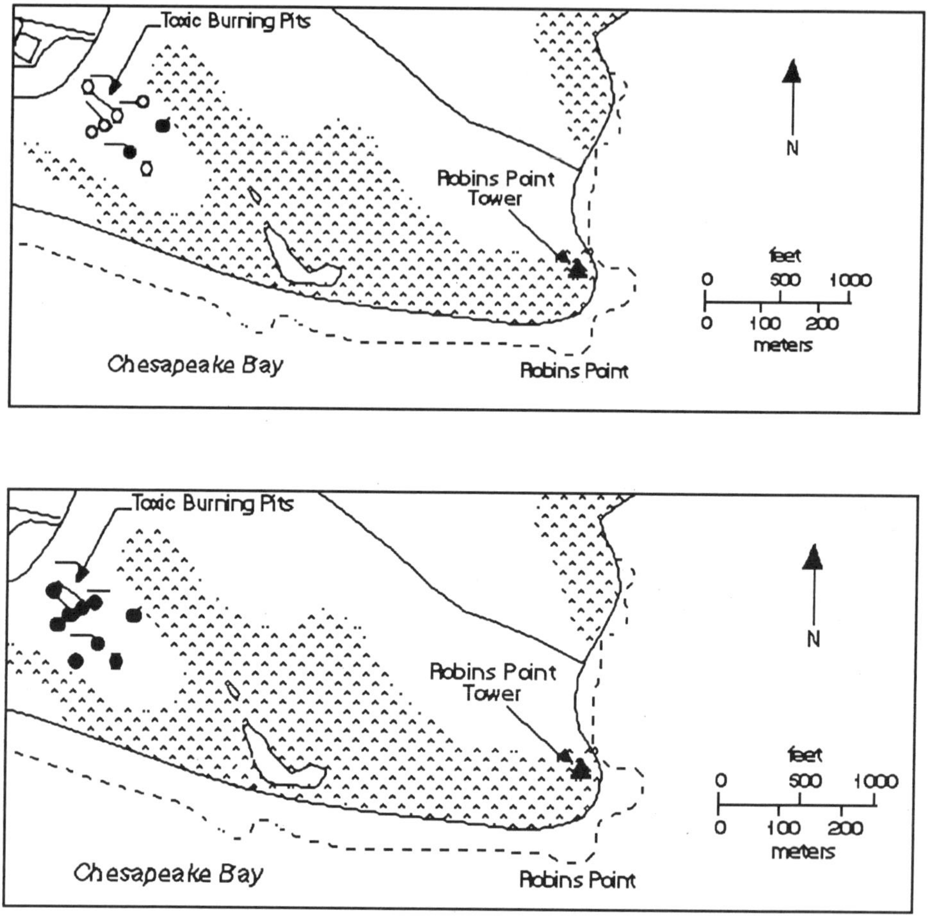

Figure 2. Earthworm toxicity screening of soils (top) and phytotoxicity screening of soils (*bottom*).

● *toxic effect;* ○ *no toxic effect*

found to be spatially correlated within themselves and negatively cross-correlated. The cross-correlation between total enzyme activity and metal concentration was analyzed using two methods: a cross-semivariogram, a standard geostatistical tool, and an interactive graphical data representation using geostatistically estimated data distributions. This estimation technique does not change sample point values and uses a least-squares minimization technique to estimate the data for display. The graphical analysis technique

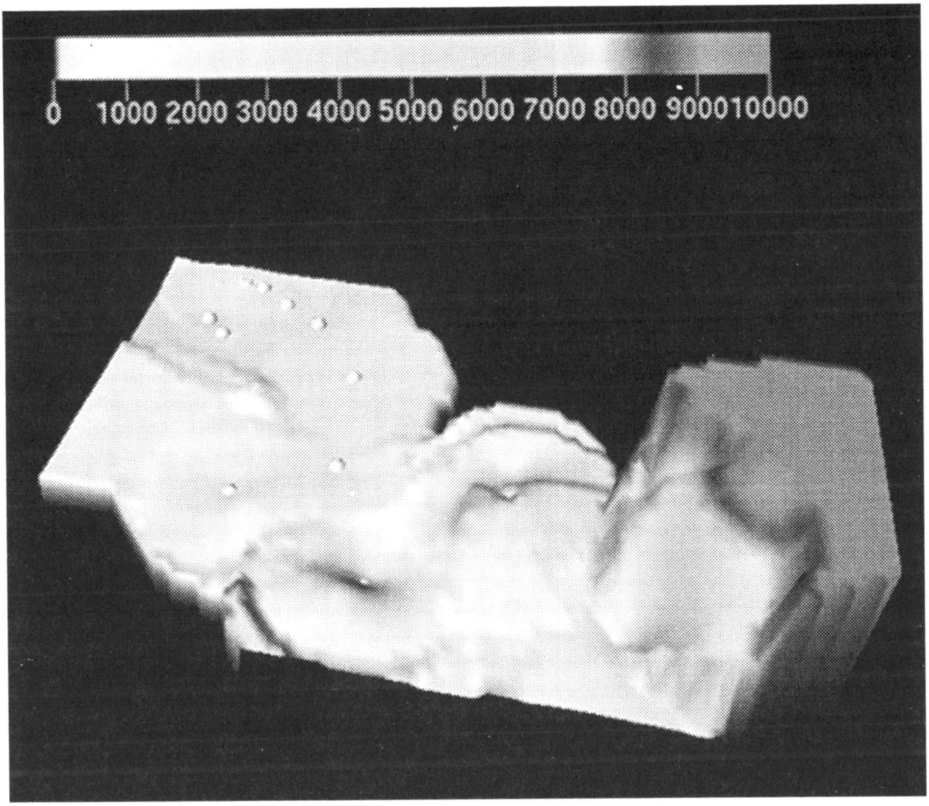

Figure 3 Spatial distribution of the sum of the concentations of Pu, Cu, and Zn.

involves displaying a surface in three-dimensional space, the topology of which is generated by one variable (total enzyme activity) and the surface colored by a second (total metal concentration). Both the visualization technique and the cross-semivariogram showed a negative correlation between the total enzyme activity and the metal concentrations.

5. Discussion

This study provided an opportunity to develop and test the efficacy of a comprehensive methodology for assessing ecological risks of soil contamination. A hierarchical approach to assessment of contaminated soils incorporated several levels of ecosystem structure and allowed ascertainment of changes in the structure and function of soil communities in an

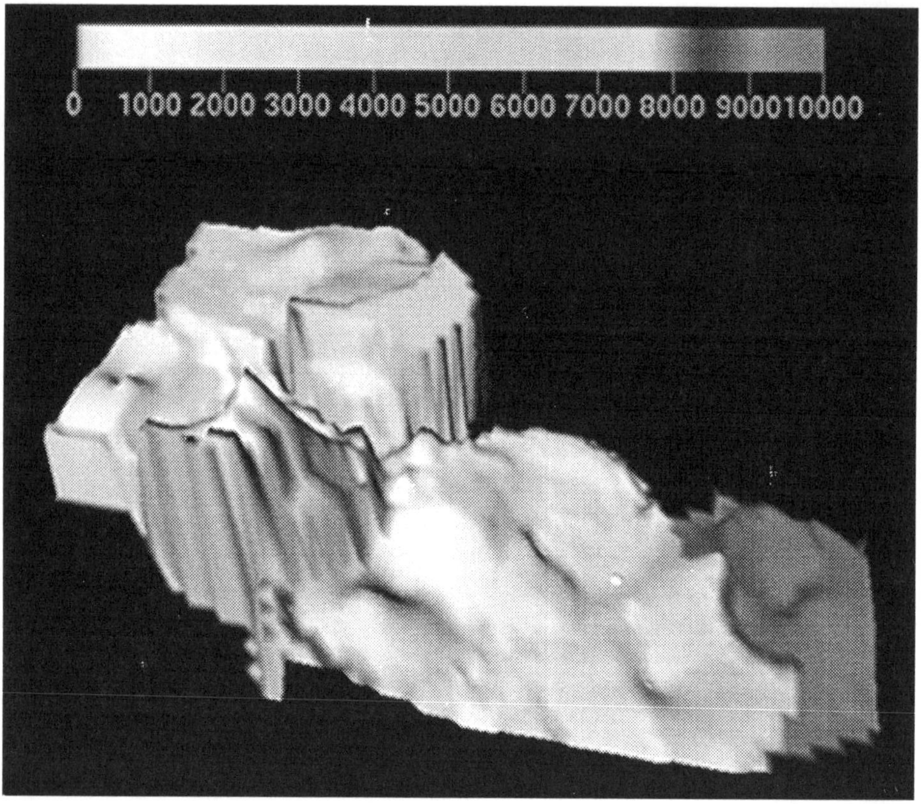

Figure 4 Spatial correlation of total enzyme activity with the sum of the concentrations of Pb, Cu, and Zn.

area with a wide range of concentrations of heavy metals. All measured population and community parameters were affected by soil contamination with heavy metals. The numbers of soil macroinvertebrates, the abundance of trophic groups of nematodes, and the biomasses of soil microorganisms were reduced in contaminated soil.

Ecosystem processes were also sensitive to changes in the soil chemical environment. Microbial activity and C and N dynamics in decomposing litter were all considerably altered by soil contamination. Significant reduction in the activities of all enzymes closely paralleled the increase in heavy metal concentrations. Ten- to fifty-fold reductions in enzyme activities were observed as heavy metal concentrations increased. These results suggest that soil contamination with heavy metals may have detrimental effects on soil biota and on the rates of organic matter degradation and the subsequent release of nutrients to aboveground communities in the area.

With few exceptions (e.g., [9]), the ecological risk assessment programs used at most Superfund sites in the United States do not include any of the soil parameters used in this study. The only two official soil toxicity tests recommended by the U.S. Environmental Protection Agency [6] and commonly used in remedial investigations include an earthworm test and a phytotoxicity test (e.g., lettuce seed germination and root elongation tests). Additionally, tests originally developed for aquatic toxicology programs, including water flea *(Daphnia)* survival, algal growth, and fathead minnow *(Pimephales)* survival tests are recommended for assessing the toxicity of soil elutrates [6]. Clearly, these tests lack ecological realism regarding their capacity to adequately determine the effects of soil contamination on the structure and function of soil communities. They can also lead to conflicting results. In our study, the conclusions from the earthworm toxicity test contradicted those obtained from the phytotoxicity tests, making it difficult to provide meaningful recommendations to decision makers.

Analyses of spatial distribution of total enzyme activity (one of the ecological response variables used in this study) allowed for the delineation of the area of J-Field where significant detrimental ecological effects occurred. These data may be useful in determining the extent of cleanup activities needed in the area of concern. This information is particularly important for Superfund sites like J-Field that do not have an unrestricted-use alternative in the future. At these sites, the guidelines for concentrations of soil contaminants based on human toxicology commonly used in remedial investigations cannot apply, and the assessment of ecological effects of soil contamination may be the only acceptable option.

Our results further indicate that a hierarchical approach can be a valuable strategy in the assessment of the ecological effects of soil contamination. By integrating biotic parameters and ecosystem processes, this approach can give insight into the mechanisms that lead to alterations in the structure and function of soil ecosystems.

Acknowledgments. I am grateful to R. Parmelee, G. Williams, M. Carreiro, E. Ingham, S. Zoars, and R. Compte for their participation in this project. This work was supported under a military interdepartmental purchase request from the U.S. Department of Defense, Directorate of Safety, Health, and Environment, Aberdeen Proving Ground, Maryland, through U.S. Department of Energy contract W-31-109-Eng-38.

6. References

1. Blair, J.M. (1988) Nitrogen, phosphorous and sulphur dynamics in decomposing deciduous leaf litter in the southern Appalachians, *Soil Biology and Biochemistry* **20**, 693–701.

2. Cheng, W., and Coleman, D.C. (1989) "A simple method for measuring CO_2 in the continued air flow modification of substrate-induced respiration technique," *Soil Biology and Biochemistry* 21:385-388.

3. Cheng, W., and Virginia, R.A. (1993) "Measurement of microbial biomass in Arctic tundra soils using fumigation-extraction and substrate-induced respiration procedures," *Soil Biology and Biochemistry* 25:135-141.

4. Crossley, D.A., Coleman, D.C., Hendrix, P.F., Cheng, W., Wright, D.H., Beare M.H., and Edwards, C.A. (1991) *Modern Techniques in Soil Ecology*, Elsevier, New York.

5. Deutsch, C.V., and Journal, A.G. (1992) *GSLIB Geostatistical Software Library and User's Guide*, Oxford University Press, New York.

6. Greene, J.C., Bartels, C.L., Warren-Hicks, W.J., Parkhurst, B.R., Linder, G.L., Peterson, S.A., and Miller, W. (1988) *Protocols for Short Term Toxicity Screening of Hazardous Waste Sites,* EPA/600/3-88/029, U.S. Environmental Protection Agency, Corvallis, Ore.

7. Ingham, E.R., and Klein, D.A. (1984) Soil fungi: relationships between hyphal activity and staining with fluorescein diacetate, *Soil Biology and Biochemistry* **16**, 273–278.

8. Karnak, R., and Hamelink, J. (1982) A standardized method for determining acute toxicity of chemicals to earthworms, *Ecotoxicology Environment Safety* **6**, 216-222.

9. Menzie, C.A., Burmaster, D.E., Freshman, J.S., and Callahan, C.A. (1992) Assessment of methods for estimating ecological risk in the terrestrial component: a case study at the Baird and McGuire Superfund site in Holbrook, Massachusetts, *Environmental Toxicology and Chemistry* **11**, 245–260.

10. Neuhauser, E.F., Loehr, R.C., Malecki, M.R., Milligan, D.L., and Durkin, P.R. (1985) The toxicity of selected organic chemicals to the earthworm *Eisenia fetida*, *Journal of Environmental Quality* **14**, 383–388.

11. Phillips, C.T., Checkai, R.T., and Stull, E.A. (1994) *Phytotoxicity and earthworm toxicity testing of soil samples from J-Field, Aberdeen Proving Ground, Maryland*, TU-7/ANL/APG/J-F/RI-ERA, Argonne National Laboratory, Argonne, Ill.

12. SAS Institute Inc. (1988) *User's Guide,* Version 6.03, SAS Institute Inc., Cary, N.C.

13. Silicon Graphics Inc. (1992) *SGI Explorer User's Manual*, Mountain View, Calif.

14. U.S. Environmental Protection Agency (1982) *Environmental Effects Test Guidelines*, EPA 560/6-82-002, Washington, D.C.

BIOPROCESSING OF ORGANIC WASTES: SAFETY FOR LANDSPREADING THEM

J. DOMÍNGUEZ and S. MATO
Department of Natural Resources and Environment
University of Vigo, SPAIN.

1. Introduction

In traditional farming systems the animal manures and slurries are widely utilized in croplands to increase soil organic matter, improve soil structure and as a source of nutrients. The incorporation of organic matter into soils also increases water holding capacity, water infiltration, and porosity. Organic matter also decreases soil crusting and bulk density, and improves cation exchange capacity.

In other hand we have reached a critical stage in the management of one of the major municipal waste stream, sewage sludge. Their disposal alternatives are basically: application to soil, dumping at sea, landfilling or incineration. With increasing pressure to ban all sludge dumping at sea and the prohibitive costs of landfilling and incineration, there is a great incentive to dispose of sludge on land, unless its composition renders it too toxic for use.

2. Adverse Effects of an Incorrect Spreading of Organic Wastes

Indiscriminate landspreading of intensively-produced animal wastes and sewage sludge damages soil fertility, causes water pollution and may present a health risk.
The potentially adverse effects of incorrect slurry application include:

SOIL:

- Elevated nutrient inputs, particularly with N, P and also with K.
- Ammonia toxicity for soil biota.
- Increasing heavy metals content in the soil top layer.
- Increasing pH values after long-term use.
- Salt accumulation in dry areas.
- Anaerobiosis and anoxic decomposition pathways.
- Spreading of bacterial, viral and parasitic diseases.
- Excessive loosening of the soil after large accumulations of organic material.

N. M. van Straalen and D. A. Krivolutsky (eds.), Bioindicator Systems for Soil Pollution, 213–221.

- Compaction of the soil after slurry application in wet soil conditions and by heavy slurry tankers.
- Ammonia volatilization contributing to acidification and N enrichment of natural areas.

WATER:
- Ammonia toxicity.
- Groundwater pollution, especially with nitrate.
- Nitrate, P, organic matter, and Cu enrichment of surface and sea waters by direct surface runoff.
- Leaching of P in moorland soils with low pH (peat soils).
- Leaching of K from sandy and limestone soils.
- Pathogen diseases.

Another type of organic wastes are those which have a high C:N ratio, such as paper-mill sludge and paper pulp-mill sludge, which pose other kinds of environmental problems when disposed of on land. These are highly stable materials with a considerable nutrient unbalance which may cause nutrient inmobilization in the soil.

3. Biological Treatments of Organic Wastes

The importance of biological processes in organic waste treatment has been thoroughly verified and is essential for their safe disposal. Within this broad range of bioprocesses, we are here concerned with two which are the most efficient in managing solid organic waste.

3.1. COMPOSTING

Composting is an accelerated aerobic biological process whereby microorganisms (mainly bacteria, fungi and actinomycetes) liberate heat, carbon dioxide and water. The heterogeneous organic material is transformed into a homogeneous and stabilized humus-like product called *compost.*[1,2]

Application of composting has been impeded by the mythology that composting is a "natural process", and control can be left to default. Successful composting requires an adequate processing control criteria.

3.2 VERMICOMPOSTING

Vermicomposting is a biooxidation and stabilization process of organic material involving the joint action of earthworms and microorganisms, by which high quality amendment or organic fertilizers called vermicompost are obtained.[3]

4. Advantages of the Biological Treatment of Organic Wastes

4.1. ADVANTAGES OF COMPOSTING

- Compost improves soil physical properties such as water-content and retention, soil aeration and permeability and water infiltration which contribute significantly to reducing soil erosion and loss of plant nutrients by runoff.
- Organic wastes can be recycled as compost, reducing the exploitation of limited resources such as inorganic fertilizers and peat, and saving energy for ammonia and nitrate production.
- Composting produces a noticeable and rapid volume and weight reduction as a consequence of organic matter degradation and losses by evaporation. In addition, nutrients content per volume and weight unit becomes higher.
- Odour problems and presence of flies and rats are removed.
- Weed seeds and pathogen organisms are eliminated by the high temperatures during the process.
- Spreading compost leads to a more uniform spreading of the plant nutrients, and unlike raw organic waste, no phytotoxic effects to seedlings and plant roots are produced.
- For seedling breeding composts represent an alternative to peat and sphagnum resources.
- Composting increases pesticide degradation.

Some organic materials of agricultural origin may be the carriers of agrochemicals such as herbicides, insecticides, fungicides and growth regulators. The composting of organic material is characterised by a very high biological activity so the composting conditions can be assumed to be very favourable for the breakdown of agrochemicals. Vogtmann *et al.*[4] examined the degradability of various agrochemicals after 56 and 75 days of composting. The tested agents were -two herbicides (Bromofenoxin, 2, 4D), four insecticides (Lindane, Diazinon, Matasystox and Parathion), two fungicides (Benomyl, Propineb) and two growth regulators one for plants (CCC) and the other for animals (Tylosin). The degradability ranged from 100% (Matasystox, Tylosin) down to 18% (CCC) whereas most of the chemicals lay over 80%. The degradation was caused by temperature and biochemical effects.

- The amounts of bioavailable heavy metals tends to decrease.

4.2 ADVANTAGES OF VERMICOMPOSTING

- Organic wastes can be rapidly broken down by earthworms, resulting in a stable non-toxic material which has an economic value as soil conditioner for plant growth.
- Vermicompost is a finely-divided peat-like material with excellent structure, porosity, aeration, drainage and moisture-holding capacity.

- Low cost systems could be developed for farming purposes.
- Vermicomposting supplies a suitable mineral balance and could act as complex-fertilizer granules.
- Vermicomposting provides a great reduction of the waste bulk density.
- Significantly lower pathogen content.
- Earthworm presence favours nitrogen mineralization.
- The amounts of bioavailable heavy metals tend to decrease.
- The final product supplies hormone-like compounds which accelerate plant growth.

5. Some Practical Cases

The following cases substantiate some of the important effects of biological treatments on organic wastes, significantly reducing their capacity to pollue and allowing them to be spread directly on the soil, not only but only on croplands but also on deteriorated areas with a view to their recovery.

5.1. VERMICOMPOSTING OF PIG SLURRY WITH AGRICULTURAL WASTES

The next tables reflect the great effect of vermicomposting on the nitrification process over different time periods. In the blank mixture, without earthworms, the nitrate formation is notably lower. The process involves a drastic reduction in ammonia content while the C:N ratio is considerably improved.

	OM	%C	%N	NO_3 (ppm)	NH_4+(ppm)	C:N
Pig slurry (beg.)	57,5	33,4	3,02	6,6	6510	11,05
Vermicompost (4 months)	55,4	32,2	2,58	154	4,4	12,48

	OM	%C	%N	NO_3 (ppm)	NH_4+(ppm)	C:N
Pig slurry (beg.)	57,5	33,4	3,02	6,6	6510	11,05
Pig slurry+straw (beg.)	70,6	41,1	3,40	4,4	4175	12,08
Vermicompost (2 months)	63,5	36,9	2,83	48	4,2	17,31
Vermicompost (4 months)	54,9	31,6	2,34	128	3,0	13,50

	OM	%C	%N	NO_3 (ppm)	NH_4+(ppm)	C:N
Pig slurry (beg.)	57,5	33,4	3,02	6,6	6510	11,05
Pig slurry+*Ulex sp.* (beg.)	64,6	41,1	3,40	4,4	4175	12,08
Vermicompost (3 months)	63,5	36,9	2,83	48	4,2	17,31
Vermicompost (6 months)	54,9	31,6	2,34	128	3,0	13,50
Without worms (3 months)	67,0	38,9	2,56	9,2	4,1	15,19

5.2. BIOASSAY OF PHYTOTOXICITY IN COMPOSTING OF SEWAGE SLUDGE WITH *Lepidium sativum*

Despite only a slight reduction in toxicity, this is significant in such a short period of the process (from 5 to 9 days). This period corresponds to the first composting stage, during which the temperature rises, without the "cooling" or long maturing stage taking place. The composting process has not been completed but it is of great interest to seek the shortest possible residence periods and which have adequately positive effects, due to the large amounts of these wastes produced.

Sewage sludge + *Ulex sp.* (9 days)

	% Germination	Toxicity index(%)
Before composting	52,5	84,0
After composting	77,5	60,7

Sewage sludge + Straw. (7 days)

	% Germination	Toxicity index(%)
Before composting	67,5	65,5
After composting	76,2	50,7

Sewage sludge + *Pteridium sp.* (5 days)

	% Germination	Toxicity index(%)
Before composting	53,7	78,2
After composting	57,5	68,6

% Germination: seedling emergence in sample/ seedling emergence in control x 100
Toxicity Index = 100 - Germination Index
Germination Index = (R_s x L_s / R_c x L_c) x 100 (Rs and Rc are the number of germinated seeds in sample and control respectively. Ls and Lc are the length of the root in sample and control respectively).
(De Bertoldi and Zucconi) [5]

5.3. HEAVY METAL CONTENT IN COMPOSTING OF SEWAGE SLUDGE

Although total metal content tends to increase after the composting process due to the breakdown of organic material and evaporation loss, the presence of these metals (measured in AB-DTPA) tends to fall. This reduction varies in each mixture which must certainly be related to the development of the fermentative process. Similar results can be observed in the Table , in the vermicomposting of paper pulp mill sludge.

Sewage sludge + Ulex sp. (9 days)

	Fe		Mn		Zn		Cu		Pb		Ni	
	Total	%Ext.	Total	%Ext.	Total	%Ext.	Total	%Ext.	Total	%Ext.	Total	%Ext
Initial	9368	14,2	359	59,1	606	27,2	212	84,3	107	51,2	35	15,8
Compost	11374	7,5	405	45	503	30,5	258	59	107	44,7	46	10,6

Sewage sludge + Straw (7 days)

	Fe		Mn		Zn		Cu		Pb		Ni	
	Total	%Ext	Total	%Ext	Total	%Ext	Total	%Ext	Total	%Ext	Total	% Ext.
Initial	9438	14,9	300	48,7	383	55,3	305	81,1	146	71,8	37	21,9
Compost	10805	10,3	372	41,3	475	47,1	334	75,7	155	51,8	40	16,6

Sewage sludge + *Pteridium sp.* (5 days)

	Fe		Mn		Zn		Cu		Pb		Ni	
	Total	% Ext.	Total	% Ext.	Total	% Ext.	Total	% Ext.	Total	% Ext.	Total	% Ext.
Initial	4701	5,7	313	75,0	344	55,9	308	78,7	86	61,7	37	9,5
Compost	4752	3,3	353	72,1	330	56,2	305	60,4	77	59,6	33	11,2

* Total values in ppm (dry weight) * % Extractibility defined as AB- DTPA metal content / Total metal content x 100.

5.4. VERMICOMPOSTING OF PAPER PULP MILL SLUDGE

The problem posed by this type of organic wastes is enterely different to that of the previous highly-nitrogenated ones. These wastes, having a high C:N rate are highly stable and are of specially slow breakdown. Therefore, deficiencies in N content in the soil may result from their spreading. Simply mixing these materials with others which are rich in N (sewage and animal slurries) becomes in itself a suitable biotreatment before their disposal on land, highly favouring their breakdown. In addition, the C:N ratio is greatly improved.

The presence of earthworms in the system may aid to complete this process, as they will accelerate the mineralization of the organic matter favouring humification processes and enabling materials of agronomic interest to be obtained.

PPMS + Pig Slurry (4:1 d.w.) (60 days)

	Ashes	TOC	TKN	C:N	pH	CF	HA	FA	NHS	TEC/TOC
PPMS	5,4	51,5	0,2	257,3	9,3	69,8	1,06	0,39	0,40	4,1
Initial	10,1	49,6	0,82	62,0	8,7	61,5	1,49	0,65	0,94	6,8
VC	15,2	41,7	2,17	19,2	7,5	26,6	3,20	1,11	4,60	21,6
Blank	19,4	44,6	2,08	21,5	7,6	27,6	3,11	1,31	4,50	21,0

PPMS + Sewage Sludge (3:1 d.w.) (30 days)

	Ashes	TOC	TKN	C:N	pH	CF	HA	FA	NHS	TEC/TOC
PPMS	5,4	51,5	0,2	257,3	9,3	69,8	1,06	0,39	0,40	4,05
Initial	25,0	42,4	1,1	40,2	8,1	52,3	1,56	0,32	1,30	7,7
VC	55,2	24,7	3,8	6,4	8,8	12,4	3,12	1,28	3,13	31,0
Blank	47,6	29,6	1,9	15,8	8,3	17,8	3,02	0,65	2,67	21,8

PPMS: paper pulp mill sludge; VC: vermicompost (*Eisenia andrei*); Blank: without earthworms; TOC: Total Organic Carbon; TKN: Total Kjeldahl Nitrogen; CF: Crude Fibre; HA: humic acids carbon; FA: fulvic acids carbon; NHS: non-humified substances carbon; TEC/TOC: total extractable carbon / TOC x 100; all data in % of dry weight.

6.3. CHANGES IN METAL EXTRACTIBILITY IN VERMICOMPOSTING OF PAPER PULP MILL SLUDGE

Although in many cases heavy metal content does not exceed recommended levels, it is very important to minimize the problems these may cause, avoiding excessive accumulation in the soil.

In diagnostic extraction for available micronutrients in soils, several chelating agents are currently used (e.g. DTPA, EDTA, AB-DTPA). Variations in heavy metal extractibility using these substances may be highly useful in the prediction of the bioavailability in organic waste added to the soil and biostabilized organic waste. The amount of metal extracted using the chelating agents tends to decrease generally during composting, vermicomposting and any maturing process of organic material.

This indicates that these elements were insoluble, possibly because of oxidation processes, which produced a number of slightly soluble oxides, hydroxides and carbonates, and because of the formation of water-insoluble complexes with the most high polymerized organic fractions.[6]

	Fe		Mn		Zn		Cu		Ni		Pb	
	Total	%Ext	Total	%Ext	Total	%Ext	Total	%Ext	Total	%Ext	Total	%Ext
pig-ini	2110	10,5	143	17,5	529	47,1	125	26,1	nd	nd	nd	nd
pig-vc	3325	7,5	272	13,6	839	55,3	219	20,8	nd	nd	nd	nd
pig-blank	3140	8,5	292	13,4	1040	49,2	265	22,8	nd	nd	nd	nd
ptry-ini	835	14,4	205	31,0	53,0	47,6	19	67,0	9	0,0	nd	nd
ptry-vc	1220	14,9	344	24,5	76	43,2	33	60,6	13	33,9	nd	nd
ss-ini	4475	12,3	165	51,8	230	66,4	189	71,5	40	24,8	68	40,4
ss-vc	6900	9,2	275	42,5	390	72,0	270	63,0	61	34,8	134	42,6

PPMS: paper pulp mill sludge; Pig: PPMS+Pig slurry(4:1 d.w.). Ptry: PPMS+Poultry slurry (3:1 d.w.). SS: PPMS+Sewage sludge (3:1 d.w.).

Total in mg.kg^{-1} on dry weight; %Ext: DTPA Extractable/ Total x 100; nd: no detectable; Cd and Co no detectables; detectability limits in mg.kg^{-1} : Ni- 0,06; Pb-0,1; Cd-0,01; Co-0,05.

7. Final Considerations

The basic aim of this project is to highlight the possibilities offered by bioprocesses towards reducing or solving ecotoxicological problems caused by spreading organic waste directly on the soil.

Composting and vermicomposting have traditionally been applied in order to obtain products which are commercially highly valuable for business ends. For this reason, the other possibilities they offer have been disregarded and left unstudied. We consider it to be of utmost importance to apply these processes in order to stabilize organic waste, while managing to solve, or at least minimize, the environmental problems arising from their disposal, without needing in many cases to complete the process.

Alternatively, if the composition and characteristics of the waste to be treated allow, the processes may be totally completed so that the end products are of better use.

8. Acknowledgements

This researh was supported by the Spanish Interministerial Commission for Science and Technology (National Plan for Biotechnology, Projects BIO89-0111 and BIO336/93).

9. References

1. Finstein, M.S., Miller, F.C., Strom, P.F., McGregor, S.T. and Psarianos K.M. (1983) Composting ecosystem management for waste treatment, *Biotechnology* **1** (4) 347-353.

2. Finstein, M.S., Miller, F.C. and Strom, P.F. (1986) Waste treatment composting as a controlled system, *Biotechnology* **8**, 363-398.
3. Edwards, C.A. (1988) Breakdown of animal, vegetable and industrial organic wastes by earthworms. In C. A. Edwards and E. F. Neuhauser (eds.), *Earthworms in Waste and Environmental Management*,SPB, The Hague, pp. 21-31.
4. Vogtmann, H., Fragstein, P. and Draeger, P. (1983) The degradation of agrochemicals during composting. Proceedings of the Second International Symposium Peat in Agriculture and Horticulture Bet Dagan, Israel.
5. De Bertoldi, M. and Zucconi, F. (1983) Annalisi della fitotossicita della sostanza organica in decomposizione mediante bioassagio *Lepidium sativum.* In *Metodi annalitici per i fanghi. Parametri biochimici e biologici*. Inst. Recerca sulle Acque CNR Roma.
6. Stevenson, F.J. (1986) *Cycles of Soil.* John Wiley & Sons, New York.

ECOTOXICOLOGICAL BIOINDICATION: PAH, PCB AND HEAVY METALS STUDIED IN THE NATURAL RESOURCE MONITORING PROGRAMME OF BERLIN (GERMANY)

WERNER KRATZ
Institute of Animal Physiology
Department of Biology, Free University of Berlin
Grunewaldstr. 34, D-12165 Berlin, Germany

Abstract: As part of the monitoring programme of Berlin and Brandenburg in a network spreading over 3.000 km^2, the accumulation of atmospheric pollutants such as polycyclic aromatic hydrocarbons (PAH), polychlorinated biphenyls (PCB) and heavy metals (Cd, Pb, Cu, Zn) were measured using bioindicator Scots Pine (*Pinus sylvestris* L.) and earthworm species. The passive monitoring technique used is based on the fact that the concentrations of PAH, PCB and heavy metals present in the atmospheric fallout are proportionally taken up and accumulated by the bioindicators. The regional background deposition levels show different distribution patterns. The PAH concentrations in the pine needles were closely related to the location of emission source, whereas PCB levels represent a more ubiquitous distribution pattern. The heavy metal contents of earthworms showed that inner city soils had higher Pb, Cd, and Cu concentrations. No regional differences for Zn burden in the earthworms was found.

Key words: emission, pollution, contamination, ecotoxicology, bioindicator, monitoring, Scots Pine, earthworm, heavy metals, lead, cadmium, copper, zinc, PCB, PAH.

1. Introduction

Berlin with over 3.5 million inhabitants and an area of 889 km^2 is one of the largest urban cities of the Federal Republic of Germany. Hence, emissions of many atmospheric pollutants from different sources such as industry, traffic, waste incineration and domestic fuel burning are high. The emission of dust and aerosols therefore require particular attention, especially for heavy metals, PAHs and PCBs contents. Polycyclic aromatic hydrocarbons (PAHs) are formed by incomplete combustion of organic material, many of which are known to be mutagenic and

N. M. van Straalen and D. A. Krivolutsky (eds.), Bioindicator Systems for Soil Pollution, 223–232.

carcinogenic. Polychlorinated biphenyls (PCBs) emitted from industrial production processes and their use as coolant are regarded as highly harmful to health because of their ability to accumulate and persist in the fatty tissues. Since 1989 the production and use of PCBs in Germany is banned. To protect the public from harmful effects of air pollutants, it was imperative to investigate the distribution of pollutants and their effects in the urban ecosystem of Berlin. Therefore in 1987 it was recommended by the expert commission on environment of Germany to initiate a state-wide and possibly country-wide ecological observation. Subsequently in 1989 and in 1990 the State Governments of Berlin and Brandenburg respectively established the Natural Resource Monitoring Programme to conduct the ecological monitoring of urban and rural areas of Berlin and Brandenburg. Therefore studies were undertaken to monitor the contamination using bioindicators from different trophic levels.

2. Aims and Objectives of the Survey

The emission ecological mapping programme established for Berlin is part of the monitoring survey of Natural Resource Monitoring Programme started since 1989, which monitors and describes regional effects of atmospheric fallout on bioindicators out of the flora and fauna. This programme covers the whole urban and the surrounding rural area of Berlin-Brandenburg and uses a monitoring grid of 48 plots at a distance of 4 km in the inner city and 16 km in its surroundings. The close density of the plots in the inner city was necessary in order to register most of the irregularly distributed single emittors. Since the pollution decreases towards the rural areas and distributes more evenly, less plots were necessary for measurements. Many of the monitoring plots were located near the already existing plots of the Berlin Air Quality Monitoring Network (BLUME) grid so that direct comparision with other data obtained was possible (cf. Fig. 1).

3. Bioindicators

Generally bioindicators which have accumulation capacity in a compartment of eco-system have the advantage that

- pollutants like heavy metals and chlorinated hydrocarbons accumulate in the organisms at a higher concentration than the surroundings in which they live, which makes chemical analysis easier,
- analysis of the organism is a measure of bioavailability of contaminants which is important to assess biological effect,
- organisms integrate contaminants over time.

The limitations of bioindicator analysis are that

- in addition to the ease of analysis, the concentration in the organisms must be a function of contamination in the environment,

- organisms analysed should be representative to allow extrapolation of results to other organisms in the system,
- there are no universal indicator organisms, therefore different organisms should be analysed for exact exposure measurements,
- fluctuation in concentration of the contaminant is sometimes of more interest as this may allow the kinetic changes and distribution in the environment and its biological effect.

As already known from other studies, pine needles directly exposed to the atmospheric pollutants are good indicators of airborne contamination and thus act as diffusive samplers. The damaging effects that occur due to pollutants are of different nature. Acute damage can be observed in areas with high pollution due to intense emission from industry and faulty industrial plants. Usually toxic effects are more difficult to notice in plants because of low concentrations and long periods of exposure. Nevertheless plants show chronic damages (necroses, chloroses) under subacute exposures. Moreover, the combination of different pollutants in relevant con-centrations often decreases the phytotoxic threshold value. In future it is recommended to consider especially these synergistic effects [1].

In the present investigation Scots Pine (*Pinus sylvestris* L.) which is a widespread tree in the Berlin area, was used as a passive monitor organism for air pollution. Because of their epicuticular wax, which consists mainly of long-chain esters, polyesters and paraffins, pine needles are considered to be good accumulators of lipophilic organic pollutants like PAHs and PCBs.

Earthworms are especially suited as soil bioindicators for heavy metals and organic pollutants. Therefore any atmospheric fallout on soil in reasonable quantity can be detected by them. They fulfill the following criteria:

- have good accumulation property for metals and organics,
- have wide distribution spectrum,
- are relevant for ecosystem processes,
- are faithful to their territory.

Little is known about the toxicological significance of the measured values of compounds and the possible transfer (indication value) to other ecosystem compartments. The following toxicity range of heavy metals in relation to mortality for *Lumbricus terrestris* L. have been shown by Ma [2]:

$$\text{lead} < \text{nickel} < \text{copper} < \text{cadmium}$$

As earthworms are of ecological significance and have a key position in the decomposition process, and in the biogenic distribution of organic matter of soil, they were used as passive monitors to investigate heavy metals present in the soil.

4. Sampling and Analysis

The pine needles were collected and monitored from 24 plot monitoring grid transect running from North-West to South-East of Berlin (cf. Fig. 1). The samples were taken in October 1991 from pine trees with one-and-half year old needles from the outer area of the tree tops. In case of group of trees, generally a sample from three trees was taken and pooled. The analysis of the samples was done on air-dried needles which were ground and extracted with n-hexane in a Soxhlet apparatus. After clean-up of the extracts on activated silica gel, PAHs were determined by HPLC and PCBs by GC.

The earthworms were collected from 48 plots of the monitoring grid at a depth of about 10 cm soil by digging and hand sorting in June 1992. All earthworm species were collected which were found in 2 to 10 m^2 area. After gut depuration the earthworms were freezed-dried, ground and analysed for cadmium, lead, copper and zinc by G-AAS. The species spectrum of earthworms in different locations were determined from the unused material at a later period. Unfortunately no soil analysis for these metals could be performed in the study regions as this would have helped in better understanding the concentration levels in earthworms.

5. Results of PAHs and PCBs Analysis in Scots Pine Needles

PAH: The six PAHs in pine needles analysed were of priority pollutant listing as of German Drinking Water Regulation. They are fluoranthene, benzo(b)fluoranthene, benzo(k)fluoranthene, benzo(a)pyrene, benzo(ghi)perylene, indeno(1,2,3-cd)pyrene. The total six PAH in pine needles show concentrations between 106 and 843 μg/kg dry weight (mean 344 μg/kg dry wt.) (Fig. 2). Fluoranthene dominates the accumulation spectrum. Its concentration is between 75 (in sampling plot 28) and 818 μg/kg (in sampling plot 16), whereas the values for benzo(ghi)perylene are below the detection limits; except in three cases (sampling plots 22, 29, 41). The concentration of indeno(1,2,3-cd)pyrene in the pine needles could only be detected in the sampling plot 29. Fluoranthene, benzo(b)- and benzo(k)fluoranthene and benzo(a)pyrene show a large regional variability. The accumulation pattern related to the city area of Berlin seems to be local and indicates existing point source emittors for these compounds (Fig. 1).

PCB: The six PCB congeners analysed in pine needles as numbered according to IUPAC were PCB 28, 52, 101, 138, 153 and 180. The composition of these congeners were suggested by Ballschmiter & Zell [3] as they are of significance in toxic waste deposit sites. The total PCB concentration in pine needles measured were between 6.6 and 45.9 μg/kg dry weight (mean 17.3 μg/kg) (Fig 3). In contrast to PAHs the six PCB congeners always exceeded the detection limit and were even in concentration. The differences in local accumulation points to the ubiquitous but heterogenous distribution of PCBs. The highest concentration of PCB in pine needles was found at the sampling plot 21 which is close to an incinaration plant (Berlin-Charlottenburg) (Fig. 3).

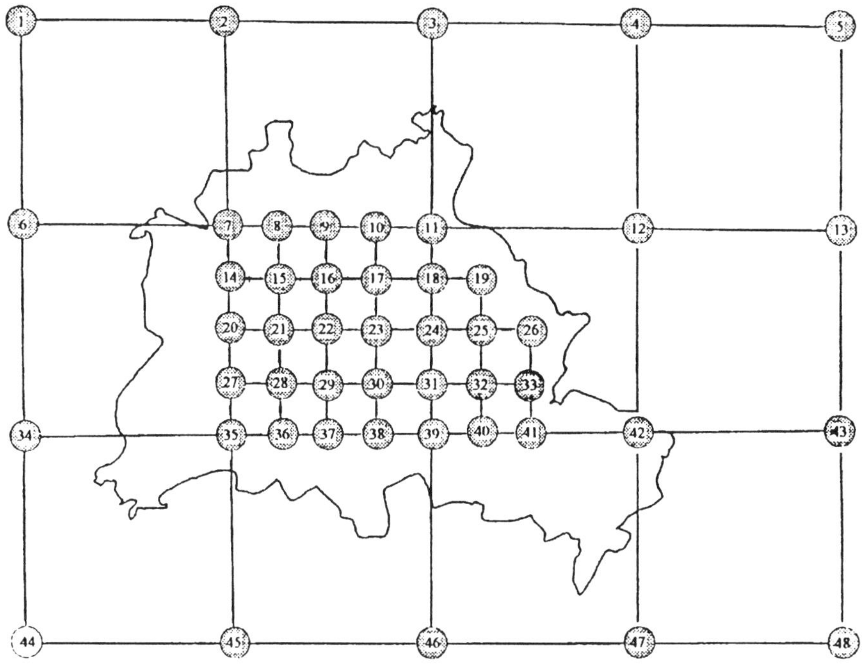

Figure 1. Monitoring grid of the Natural Resource Monitoring Programme of Berlin-Brandenburg

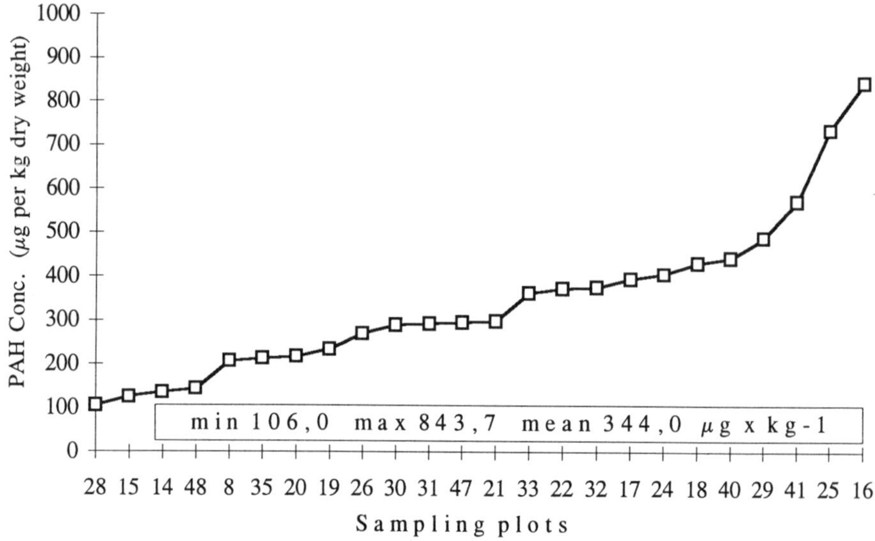

Figure 2. Σ PAH concentrations in Scots Pine needles in different plots of the Natural Resource Monitoring Programme Berlin-Brandenburg (sampling period October 1991)

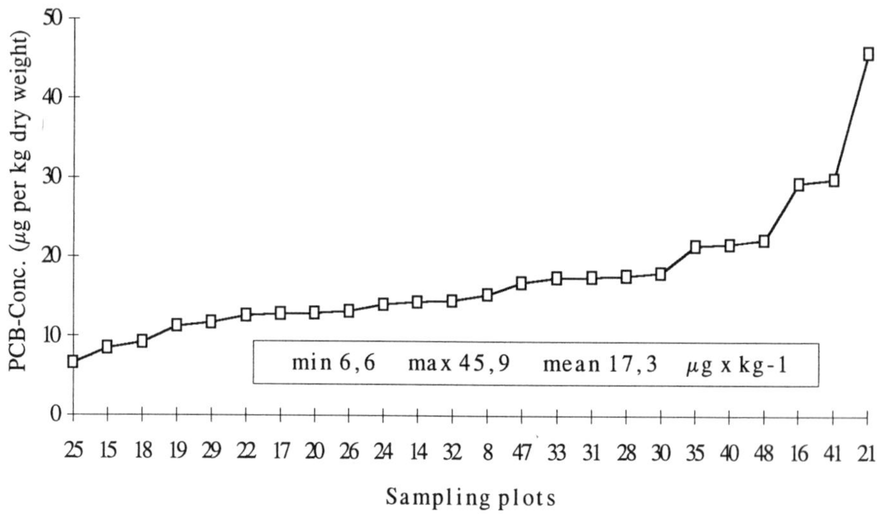

Figure 3. Σ PCB concentrations in Scots Pine needles in different plots of the Natural Resource Monitoring Programme Berlin-Brandenburg (sampling period October 1991)

6. Results of Heavy Metal Analysis in Earthworms

Lead: The measured lead concentration in earthworms ranged between 0.2 to 120.6 μg/g dry weight. When values are distributed in contamination classes then 22 sampling plots are classified as uncontaminated (normal) and the plot 17 as highly contaminated (Tab. 1). The rest of the sampling plots are of average (low and middle) contamination. This regional distribution pattern of classification appears to be vague and does not show clear zonations. By evaluating the lead values in a different scale Kratz [4] arrived at different conclusion than the LfU [5] which is responsible for the monitoring programme in Baden-Württemberg (Germany). When measured values are divided into three alike classes (e.g. 33 1/3 quantil) to show regional differences, then the North-East city area (mainly industrial activities) shows higher contamination values (this includes the plots 9, 10, 16-19, 22, 25, 29-32)

TABLE 1: Contamination values of lead in earthworms (μg/g dry weight)

Value range	Evaluation Kratz [4]	n*	Value range	Evaluation after LfU [5]	n*	33 1/3-quantil value range	n*
0 - 5	normal	22	0 - 10	normal	35	0 - 3.6	16
5 - 10	low	13					
10 - 75	middle	12	10 - 75	middle	12	3.6 - 8.3	16
> 75	high	1	> 75	high	1	8.4 - 121.0	16

n*: number of investigated plots present in the value range

Cadmium: The cadmium contamination values of earthworms ranged from 0.3 (plot 8) to 36.7 μg/g dry weight (plot 17). Because of high toxicity of this element, the values have to be seen more critically than those of lead. The contamination level of earthworms on 18 plots was classified as middle and on 22 plots as highly contaminated (Tab. 2). The regional distribution pattern of cadmium show South-East contamination in the city plots 17-19, 23, 25, 30, 32, 33, 40, and 41. As for lead contamination there is an island in plot 24 with middle cadmium contamination, in plot 17 with high contamination and in plot 31 with low contamination levels. The surrounding rural regions can not be seem as reference areas, as there are two plots (plot 5 and 6) with high contamination levels and the rest of the plots show heterogenous middle and low contamination. The northern regions tend to show low contamination levels (plot 1, 2, 3, 4).

TABLE 2: Contamination values of cadmium in earthworms (μg/g dry weight)

Value range	Evaluation Kratz [4]	n*	Value range	Evaluation after LfU [5]	n*	33 1/3-quantil value range	n*
> 0.5	normal	1					
0.5 - 5.0	low	7	0 - 10	low	18	0 - 9.5	16
5.0 - 15.0	middle	18	10 - 15	middle	8	9.5 - 18.0	16
> 15.0	high	22	> 15	high	22	18.0 - 37.0	16

n*: number of investigated plots present in the value range

Copper: The copper contamination values of earthworms ranged from 3.1 (plot 8) to 35.4 μg/g dry weight (plot 32), which is like the concentration range for cadmium (cf Tab. 2). The contamination levels in 6 plots showed low, in 37 plots as middle and in 5 plots as high (Tab. 3). The regional distribution pattern for copper also showed the same strong classes (33 1/3 quantil). Thus in the western regional area (plot 1, 6, 34. 44, 45) there was middle to high contamination levels, whereas in the rest of the region (plot 2-4, 12, 13, 42, 43, 46-48) there were low contamination levels.

TABLE 3: Contamination values of copper in earthworms (μg/g dry weight)

Value range	Evaluation Kratz [4]	n*	33 1/3-quantil value range	n*
> 5	normal	0		
3 - 10	low	6	3 - 14	16
10 - 25	middle	37	14 - 19	16
> 25	high	5	19 - 36	16

n*: number of investigated plots present in the value range

Zinc: The zinc contamination values of earthworms ranged from 22 (plot 8) to 1,766 μg/g dry weight (plot 36) and has a high variability (Tab. 4). 30 plots are seen as uncontaminated (62.5%), whereas 12 plots as middle contaminated and 6 plots as highly contaminated. No regional contamination pattern for zinc can be determined.

TABLE 4: Contamination values of zinc in earthworms (μg/g dry weight)

Value range	Evaluation Kratz [4]	n*	33 1/3-quantil value range	n*
< 500	normal	30	20 - 300	16
500 - 1000	middle	12	300 - 520	16
> 1000	high	6	520 - 1770	16

n*: number of investigated plots present in the value range

TABLE 5: Comparison of concentration range of heavy metals in earthworms (μg/g dry weight)

	present study			Weisenfeld [6] Berlin		Kratz *et al.*	LfU [5]		Martin & Coughtrey [8]	
element	min	max	med	Dahlem	Rudow	[7]	min	max	min	max
lead	0.2	120.6	10.9	15.0	60.0	85.0	1.5	958	0.3	7593
cadmium	0.3	36.7	14.8	20.0	88.0	10.3	2.6	30	0.2	144
copper	3.1	35.4	16.8	n.d.	n.d.	n.d.	n.d.	n.d.	0.2	46
zinc	22	1766	514	n.d.	n.d.	n.d.	n.d.	n.d.	75	1950

7. Discussion

On the basis of data obtained it is not possible to evaluate heavy metal concentration in earthworms because of the absence of basic ecotoxicological guide values. However comparing the concentration values for cadmium, lead, copper and zinc with other investigations, it turn out that heavy metal concentrations of earthworms in the Berlin area are in the same contamination range (Tab. 5). The integrated evaluation of the data shows that all the heavy metals determined are high in plots 17, 25, 27 and 32, whereas in plots 8 and 13 the values are generally low. The contamination burden of the surrounding rural region is mainly heterogenous, where the southern region (plots 44 - 47) has low contamination level and shows constant low levels (Fig. 1).

The regions with high percentage content of copper are incidently also with absolute low copper, lead and cadmium contents (plots 1, 2, 4, 8, 13, 20). This suggests that lead, cadmium and copper are present in non-contaminated natural regions, where copper naturally is deficient in these regions. The problem with copper and zinc is that earthworms can regulate these elements so that bioaccumulation is strongly dependent on the physiological condition of worms [9]. Besides this, copper is an essential element for earthworms which is required for specific biochemical processes.

Pine needles are good indicators of PAH accumulation as seen comparing the PAH values of airborne particles determined in other studies for the same regions [10]. Fluoranthene was the dominant component in the PAHs pollution spectrum. This substance appears to have a high photolysis stability [11] and is considered to be the main component that remains deposited on the pine needles. Because of this property and its relevant role in atmospheric pollution, it is suggested to confine on fluoranthene analysis from the PAHs spectrum in future studies. The mean fluoranthene value obtained in airborne particles during May 1990 to October 1991 from the same sampling plots (BLUME) was 5.4 μg/g dry weight [10]. However, pine needles exposed for 1.5 year period in the present study gave a mean value of 299.3 μg/kg dry weight of fluoranthene. This suggests that pine needles are more powerful bioindicators than other physical methods employed to monitor air quality.

At present ecotoxicological assessment of PAHs levels in pine needles is not possible because of the lack of basic data. The PAHs concentrations found in the Berlin monitoring programme can be compared to the Swedish study on pine needles [12]. The Swedish study found a mean fluoranthene concentration of 10.0 μg/kg dry weight. The mean value in the present study from all the plots measured accounted 344.0 μg/kg dry weight, which suggests that fluoranthene concentration in the Berlin region in 1991 was higher by a factor of 34.4.

The concentration of PCBs in pine needles was compared with the PCB values obtained in the European Monitoring Programme. The study by Jensen *et al.* [13] found PCB mean concentration of 6.2 μg/kg dry weight in the period 1984 to 1986. The values found in Berlin study are higher by a factor between 1.4 and 7.4. But it should be noted that Jensen *et al.* [13] collected their samples from remote areas. Besides, they measured the whole range of PCB congeners where the accumulated concentration may change with time [14]. However, compared with earlier studies the concentration of PCBs in pine needles have increased. As basic guide values are absent, no ecotoxicological assessment of PCBs accumulation are possible at present.

8. Conclusion

Pine needles and earthworms are suited as passive accumulators of biological system in biomonitoring air pollutants. However, it is not possible at present to correlate between the periodic contamination and the measured pollutant. Although it is suggested in the literature that the investigated organic pollutants because of their lipophilic character can indicate the constant pollutant levels (environment/plant cuticula), which depend on the air pollution during the growth of pine needles.
The heavy metal (Cd, Pb, Cu, Zn) accumulation by earthworms correlates well with the soil physical and chemical parameters, as reported in the literature, depending on the elemental speciation. For future work it is suggested to analyze only Pb and Cd in earthworms and the soil environment for monitoring purpose.

9. Acknowledgement

The present research was financially supported by the Senate of Berlin. The author thank Dr. A. Haque and H. Lohner for reviewing and copreparing the english manuscript. The views expressed herein by the author do not necessarily represent the opinions, policies, or recommendations of the Senate of Berlin.

10. References

1. Umweltbundesamt (eds.) (1988) *Luftreinhaltung ´88, Tendenzen - Probleme - Lösungen,* Erich Schmidt-Verlag, Berlin.
2. Ma, W.C. (1982) The influence of soil properties and worm related factors on the concentration of heavy metals in earthworms, *Pedobiologia* **24**, 109-119.
3. Ballschmiter, K. and Zell, M. (1980) Analysis of Polychlorinated Biphenyls by glass capillary gas chromatography; Composition of technical Aroclor and Clophen-B mixtures, *Fresenius Z. Anal. Chemie* **302,** 20-28.
4. Kratz, W. (1992) *Passives Monitoring zur Schwermetallbelastung von Regenwürmern im Monitoringprogramm Naturhaushalt Berlin und Umland,* Report - Senatsverwaltung für Stadtentwicklung und Umweltschutz, Berlin.
5. Landesanstalt für Umweltschutz Baden-Württemberg (LfU) (1988) *Immissionsökologisches Wirkungskataster Baden-Württemberg, Jahresbericht 1987,* Karlsruhe.
6. Weisenfeld, P. (1989) Cadmium- und Bleibelastung von Regenwürmern industrienaher Kleingärten und ähnlicher Bereiche in Berlin (West), *Zool. Beitr. N.F.* **32**, 301-320.
7. Kratz, W., Bielitz, K., Jaeger-Volmer, J., Kielhorn, U., Kronshage, J., Rink, U. and Weigmann, G. (1989) Ökosystemare Untersuchungen zur Schwermetallkontamination eines immissionsbelasteten Forstsaumes an der Autobahn AVUS in Berlin (West), *Verh. Ges. Ökol.* **XVIII**, 409-413.
8. Martin, M.H. and Coughtrey, P.J. (1982) *Biological monitoring of heavy metal pollution*, Applied Science Publishers, London, New York.
9. Streit, B., Krüger, C., Lahner, G., Kirsch, S., Hauser, G. and Diehl, B. (1990) Aufnahme und Speicherung von Schwermetallen durch Regenwürmer in verschiedenen Böden, *UWSF - Z. Umweltchem. Ökotox.* **2 (1)**, 10-13.
10. TÜV (1991) *Immissionsmeßprogramm Berlin 1991*, Berlin.
11. Koch R. (1989) *Umweltchemikalien*, VCH Verlag, Weinheim.
12. Thomas W., Rühling Å. and Simon H. (1984) Accumulation of airborne pollutants (PAH, chlorinated hydrocarbons, heavy metals) in various plant species and humus, *Environ. Pollut., Ser. A,* **36**, 295-300.
13. Jensen, S., Errikson G. and Kylin H. (1992) Atmospheric pollution by persistent organic compounds: Monitoring with pine needles, *Chemosphere* **24**, 229-245.
14. Kylin, H., Grimvall, E. and Östman, C. (1994) Environmental monitoring of polychlorinated biphenyls using pine needles as passive samplers, *Environ. Sci. Technol.* **28**, 1320-1324.

HEAVY METALS IN RUSSIAN WETLANDS

A.V. ZHULIDOV
Hydrochemical Institute
Federal Service of Russia for Hydrometeorology and Environmental Monitoring
Stachki Avenue 198
Rostov-on-Don 344104
Russian Federation

1. Introduction

Anthropogenic heavy metal pollution is the most important factor determining the quality of Russian wetlands. The amount of heavy metals introduced to the surface of wetlands with atmospheric precipitation has been about 1000 kg/km^2/year in the areas of ferrous and non-ferrous metallurgy and petroleum refining industry [1]. In 1993 the amount of waste discharged into Russian surface waters was 68.2 km^3, 40% of this amount being considered as untreated or inadequately treated [2]. A major amount of waste water is discharged into Russian surface waters by industries (37%) and municipal treatment plants (45%); the amount of waste discharged to Russian surface waters from 1989 to 1993 actually has not changed and has been 27-28.0 km^3 [2]. Today, 80% of solid waste in Russia is stored without necessary technological safety measures. Solid waste of the metallurgic industry in Russia is 50-80 mln tons per year. Metal plating shops in Russia contribute about 500,000 tons of waste per year. Liquid waste of these shops are discharged to surface waters or municipal channelization actually without pretreatment [1].

A natural heterogeneity in the biogeochemical background of heavy metal concentrations in landscapes is typical for the former Soviet Union territory. This has caused the development of "natural" hot spots of endemic diseases among the local population and domestic animals. Anthropogenic pollution of the environment with metals in a number of regions comes on top of the natural variability in background concentrations, and this causes the formation of areas with extreme living conditions [1, 6].

N. M. van Straalen and D. A. Krivolutsky (eds.), Bioindicator Systems for Soil Pollution, 233–247.

2. Brief classification of Russian wetlands

Wetlands are ecosystems that develop at the water - land interface, where water, bottom sediments, hydric soils, and biota interact. Wetlands can often be found along marine and oceanic coasts, in the areas adjacent to water bodies and streams, relief depressions, and groundwater outlets. A number of terms are used in Russia to designate wetlands. Both common terms are used (for instance: wet meadows, bogs, swamps, coasts, *etc.*), as well as geographically localized toponyms (laidy, alasy, mari, plavni, sory, tugai, zaimischa, kardashi, iskari, uremy, grass viski, sogry, sazy, ryamy, olsy, vatty, badarany, *etc.*). Sometimes, one and the same term is understood differently in different regions of Russia. For example, the term "laidy" designates both meadows at marine coasts and lakes in the Vasuganie (interfluve of the Irtysh and Ob rivers, South-West Siberia) and depressed waterflood sections in the tundra [3, 4, 6].

The transitional character of wetlands (in the sense that they can be considered both as terrestrial and aquatic ecosystems) makes them diverse and specific. From a biogeochemical point of view, wetlands, compared to deep-water habitats, are characterized by:

- lower pH values which are more variable in space
- lower dissolved oxygen concentrations
- extreme reducing conditions
- more favorable conditions for photo- and biodegradation, chelation and formation of organic and organomineral complexes, and higher sulphide concentrations.

Water is, however, the major factor for the existence of wetlands, irrespective of their diverse character: they are either covered by water or temporarily flooded, or saturated with water. According to B.B. Polynov's classification, the Russian wetlands are designated as aquatic or super-aquatic elementary landscapes [3, 6].

The wetland classification developed by the U.S. Fish and Wildlife Service [5] has been tested in the former Soviet Union since 1988 with the purpose of developing a unified international wetland taxonomy. The experience of this work revealed the efficiency of the U.S. FWS classification for wetland taxonomy in the former Soviet Union (Russia). Certain additions are, however, necessary at the level of subclasses. Palustrine wetlands were the only exception: the U.S. FWS classification turned to be inappropriate for these ecosystems [3, 6].

Today in Russia, the term "wetland" (in accordance with the U.S. FWS classification) is used to designate a humid habitat covered by a thin water layer or occasionally flooded or saturated with water. The water table in a wetland is either at the surface or close to the soil (ground) surface. An ecosystem may be considered as a wetland if it possesses at least one of the three following features [3, 5, 6]:

- The area is periodically or permanently inhabited by hydrophytes;

- The ground surface is not drained well and hydric soils may form in certain conditions;

- Hydric soils (or their equivalent) are saturated with water or covered by it during a vegetation period every year; the hydraulic regime may vary from permanent flooding or saturation to temporal water at the land surface. Hydric soils wil remain humid to the extent that anaerobic conditions may develop temporarily.

Hydrophytes may be represented by any plant species inhabiting the water or the ground surface (hydric soil if any), where anaerobic conditions or conditions close to anaeroby may develop (at least temporarily) due to high humidity.

It must be stressed that wetland ground should not necessarily be represented by some hydric soil (for instance, there is no hydric soil at pebble and sand beaches) but the substrate should be saturated with water or flooded at least for some time during a vegetation period [3, 6].

Inland boundaries of marine, estuarine, riverine, and lacustrine wetland ecosystems are determined by minimum and maximum water levels during a vegetation period. The boundaries between typical terrestrial ecosystems and wetlands are the areas where mesophytic and xerophytic vegetation replaces the hydrophytic one. In addition, such areas are characterized by certain variations in the soil hydraulic regime from temporal flooding (or saturation with water) during a vegetation period every year to the absence (or irregularity) of flood events (or saturations). The boundaries between riverine and lacustrine wetlands and deepwater habitats are determined by the distribution of emergent vascular plants (for lacustrine wetlands this is the area with a depth of the water column up to 2 m). The area further away is designated as a deepwater habitat [3, 5], and these habitats (deeper than 2 m for lakes) are not considered as wetlands. Their primary feature is that water, not air, is the medium inhabited by dominating organisms irrespective of the fact whether they are attached to an underlying substrate or not. In this case, the underlying substrate is not a soil [6].

The boundary between riverine and estuarine wetlands is determined by the concentration of halogene salts which is 0.05 promille; between riverine and lacustrine wetlands it is determined by the prolongation of the lake shore line; palustrine wetlands are often limited by any of the four other wetland systems (riverine, lacustrine, marine, and estuarine) [6].

3. Sampling and analysis of metals in wetland compartments

Two approaches were applied to develop information on the status of wetlands in the territory of the former Soviet Union and Russia and to collect samples for metal analysis in wetland compartments:

- Field surveys of wetlands in various regions and collection of samples for laboratory analysis of metal concentrations [3, 6, 7, 8]; some samples received from other specialists and institutions were analyzed as well.

- Analysis of information received by the Hydrochemical Institute from regional stations (Administrations) within the USSR Federal Service of Observations and Control of Environmental Pollution (OGSNK) and information that is collected presently by the Russian Federal Service of Observations and Control of Environmental Pollution (GSN).

The OGSNK was erected in the former Soviet Union in 1973, based on the network of the Hydromet Service stations. By the end of the 1990s, the OGSNK carried out observations of surface freshwaters, measuring physical, chemical, and biological parameters with simultaneous estimation of hydraulic characteristics in control sites [1, 8]. The observations (water chemistry and, to lesser extent, aquatic biology parameters) covered 1949 rivers and 287 lakes. Primary treatment of the samples for water chemistry and aquatic biology parameters was carried out in 130 and 30 laboratories, respectively. The information obtained was then presented to the Hydrochemical Institute to be included in a special data bank [6].

Today, monitoring of the Russian inland surface water pollution is done by the GSN for physical, chemical, and biological (including toxicological) parameters. Physical and chemical parameters are recorded in 1326 water bodies and streams (rivers, lakes, and reservoirs), with 1892 sites of observation. Analyses of water samples (about one million samples per year), collected by the GSN, were carried out in 80 laboratories, 87% of the analyses were carried out within the framework of a programme for regional observations, 6% of the analyses were carried out to provide accuracy control of the measurements, and 7% of the analyses were additional orders. The measurement of biological parameters by the OGSNK was started in the end of the 1970s; toxicological parameters have been measured by the GSN since 1993. The latter measurements were conducted at about 10% of all the GSN observation sites [4, 6]. The overall density of the GSN network (the number of sites per square kilometer) varies in different regions of Russia, from 0.02 (North-East Siberia and Central Siberia) to 0.8 (central part of the Russian Plane) [6].

Sampling equipment made of metal-free material was used to collect samples of wetland compartments for biogeochemical analysis. Water samples were filtered through 0.4 μm nuclear filters (Nucleapore). In the former Soviet Union, 0.7 μm cellulose-nitrate filters (Mytischi), 0.5 μm polyethylene terephthalate filters (Dubna), 0.45 μm Millipore cellulose acetate-nitrate filters and other filters including paper have all been used. Application of all these different filters gave inadequately high results of metal concentrations in water [1]. While generalizing the information the authors disregarded the results of such analyses, except for data referring to extremely polluted ecosystems, based on the information obtained from the OGSNK and GSN network of permanent stations. The most reliable data were used for compilation of the tables. These data were obtained as a result of specific investigations conducted by specialists of the Hydrochemical Institute, usually independent specialists, working on their theses, or (Soviet) Russian specialists working in cooperation with foreign scientists in the framework of international projects, and specific inter-institutional research programs in the former Soviet Union.

In case water samples were preserved, preservation was done immediately after filtration followed by acidification with nitric acid to pH < 2. When mercury was measured, and shipment of the samples took more than 6 days, an amount of 5% $K_2Cr_2O_7$ was added to the filtrates to a concentration of 0.05%.

Ashing of samples of suspended solids, bottom sediments, hydric soils, and peat was performed with acid and nitric acid fumes in a Rinse apparatus [3, 7, 8].

Analysis of metals in environmental samples was mainly conducted with atomic absorption techniques [7, 8, 9]. Colorimetry and neutron activation were used in a number of cases [7, 8].

Analysis of the information on long-term dynamics of metal concentrations in the water of different Russian wetlands showed a 56-, 66-, and 355-fold decrease in Cd, Pb, and Zn concentrations in the Usman River (Voronezh Biosphere Reserve) during an 18 year period, from 1975 to 1993 [1]. A similar change was noted in dissolved zinc concentrations at the Vashutkiny Ozera wetlands (the Bolshezemelskaya tundra): 1978: 4.2 μg/L, 1983: 2.1 μg/L, 1987: 1.5 μg/L, 1992: 0.04 μg/L, and 1994: 0.06 μg/L [1, 4, 6]. This drop in metal concentrations in the Usman River and the Vashutkiny Ozera water was related to a decrease in the extent of contamination of samples during sampling, filtration, preservation, shipment, and analysis [1]. The same cause explained a tendency of decreasing metal concentrations in water of other rivers during the period from 1958/59 to 1988 [10] and lead in Lake Huron water from 1965 to 1980 [11].

Data on long-term trends of metal concentrations in water in the Usman River and Vashutkiny Ozera wetland water showed the seriousness of errors in studies of metal concentrations in natural waters in the former Soviet Union. Thus, when analyzing data on metal concentrations in water in baseline areas of Russia, one should be very careful in using the information published in the scientific literature [1, 4, 6].

4. Geographic variability of heavy metal concentrations in wetlands within the territory of Russia

A biogeochemical classification of the regions of the former Soviet Union territory was developed by V.V. Kowalsky and coworkers [12, 13]. In conformity with wetlands, these investigations were supplemented by A.V. Zhulidov and E.M. Emetz [1, 4, 14].

The territory of Russia is divided into biogeochemical zones (regions) [1]. Within these major regions, so-called zonal and azonal biogeochemical provinces are distinguished. Azonal provinces arise from landscapes stretching from west to east as a distinct band, developed over ore bodies in areas of volcanic activity, or lacking run-off. Unlike zonal provinces, the biogeochemical parameters of the ecosystems in azonal provinces do not correspond to the parameters of the biogeochemical zone in which they are located. A detailed description of biogeochemical zones (regions) of the territory of the former Soviet Union and Russia is presented in the works of A.V. Zhulidov and V.M. Emetz [1, 14].

Measured concentrations of some heavy metals (Hg, Cd, Pb, Zn, and Cu) and their accumulation coefficients (hydric soils and bottom sediments) in invertebrates of 147 taxons of the Russian background wetlands are presented in the book of A.M. Nikanorov, A.V. Zhulidov and E.M. Emetz [3]. Ecological factors determining heavy metal concentrations in soil and wetland organisms in Russia are considered in detail in specific publications [3, 8, 15].

5. Extremely polluted freshwater bodies in Russia

The examples describing hydrographic areas were selected among all the diversity of Russian wetlands exposed to anthropogenic heavy metal pollution. The priority was given to those water bodies and streams whose condition was controlled by the OGSNK [16, 17]. Additional results of investigations carried out by the author as well as the materials of the reviews compiled by the Hydrochemical Institute were used [6].

Measured concentrations of some heavy metals (Hg, Cd, Pb, Zn, and Cu) and their accumulation coefficients (hydric soils and bottom sediments) in invertebrates of 147 taxons from Russian wetlands exposed to anthropogenic impact are presented in [8].

5.1. THE BALTIC HYDROGRAPHIC AREA

Wetlands of the Baltic hydrographic area within the former Soviet Union are polluted primarily by the effluents of chemical, pharmaceutical, forest, wood-working, food, and coal mining industries. Natural waters are exposed to significant impacts of leather factories, breweries, and pulp mills as well as municipal sewages and agricultural run-off [6]. By way of example, we provide some information on the Pregola River wetland in the city of Kaliningrad, Russia.

The principal pollution sources at this location are the effluents of ferrous metallurgy, pulp mills, ship building, electronic, and machine-building industries of the cities of Chernyakhovsk, Gvardeysk, and Kaliningrad (Pregola River). Zinc is the principal metal pollutant; concentrations were measured as follows [6, with additions]:

Water: 75 ± 9 μg/L
Seston: 360 ± 56 μg/g dry weight
Bottom sediments: 1600 ± 181 μg/g dry weight
Periphyton (*Ulotrix zonata*, *Cladophora glomerata*, *Sprirogira* sp., *Anabaena constricta*, *Navicula gracilis*, *N. rhynchocephala*, *Cocconeis pediculus*, *C. placentula*, *Nitzschia acicularis*, plus bacterial film): 950 ± 91 μg/g dry weight
Hydric soils: 2100 ± 450 μg/g dry weight
Lumbricidae (without division into taxa): 3900 ± 430 μg/g dry weight.

The frequency of suppressed periphyton (due to various zoogley, filamentous, bacilliform, cocci, spiral, and other forms of bacteria and fungi) was 50 to 75%, based on a visual scale. The phytoplankton composition downstream the city of Kaliningrad was dominated by the α-saprobic blue-green algae of the genus *Oscillatoria* with a density varying from 2.6 to 3.8 cells/mL (up to 60% of the total number) [6].

5.2. THE BLACK SEA HYDROGRAPHIC AREA

Wetlands of the Black Sea hydrographic area within the former Soviet Union are mainly polluted by organic pollutants which results in eutrophication of the water. The principal sources of pollution are the effluents of major industries, animal waste, industrial and municipal sewage, uncontrolled sewage of settlements, as well as the farm-land run-off [6].

For example, the Seim River wetlands, close to the city of Kursk, Russia, are polluted by heavy metals. Metal concentrations in the riverine wetlands are determined by industrial effluents of the city. Zinc is principal metal. Zinc concentrations in the period 1981 to 1993 were as follows [6, with additions]:

Water: 21 ± 4 μg/L
Seston: 140 ± 12 μg/g dry weight
Bottom sediment: 800 ± 91 μg/g dry weight
Periphyton (*Cladophora glomerata*, *Cymbella hervetica*, *Melosira granulata*, *Diatoma vulgare*, *Nitzschia palea*, *Cocconeis placentula*, bacterial film): 400 ± 47 μg/g dry weight

Lumbricidae (without division into taxa): 2100 ± 320 μg/g dry weight.

The frequency of suppressed occurrence of periphyton (see above) was 30 to 50% [6].

5.4. THE AZOV HYDROGRAPHIC AREA

Wetlands of this area within the boundaries of the former Soviet Union are polluted by the effluents of machine building, foods, and chemical industries, municipal wastes, agricultural run-off and animal wastes, mine waters and subshale waters of river ships [6]. As an example, we choose the Temernik River wetlands in the city of Rostov-on-Don, Russia. Heavy metal concentrations in the riverine wetlands are determined by the city sewages. Zinc is the principal metal pollutant. Zinc concentrations were measured as follows [6, with additions]:

Water: 80 ± 10 μg/L

Seston: 450 ± 51 μg/g dry weight,

Bottom sediments: 1600 ± 170 μg/g dry weight

Periphyton (*Cladophora glomerata*, *Navicula gracilis*, *Nitzschia palea*, *Synedra ulna*, *Diatoma vulgare*, *Phoicosphenia curvata*, bacterial film): 800 ± 103 μg/g dry weight

Lumbricidae (without division into taxons): 2100 ± 320 μg/g dry weight.

The frequency of reduced occurrence of periphyton (see above) was 50 to 75% [6].

5.4. THE BARENTZ SEA HYDROGRAPHIC AREA

Wetlands of the Barentz Sea hydrographic area are polluted with effluents of non-ferrous and ferrous metallurgy, chemical gas, oil refining, pulp mill, mining, and coal industries, energy and municipal economies. The total amount of waste waters received by water bodies of Murmansk oblast (The Kola Peninsula, Russia) in 1985 was more than 2 billion cubic meters [17].

The Nuduay River wetlands in the city of Monchegorsk, Russia, are probably the most heavily polluted wetlands of the Kola Peninsula. The principal source of pollution is the effluent of the "Severonickel" company. Copper and nickel are the principle metal pollutants. Copper concentrations in the Nuduay River wetland were as follows [6, with additions]:

Water: 82 ± 10 μg/L (based on the 1985 OGSNK data; the annual mean concentration in river water in the Monchegorsk area was 144 μg/L [17])

Seston: 210 ± 15 μg/g dry weight

Bottom sediments: 730 ± 88 μg/g dry weight

Periphyton (*Cladophora glomerata*, *Oscillatoria* and *Cymbella ventricosa* species, bacterial film): 400 ± 37 μg/g dry weight

Lumbricidae (without division into taxa): 320 ± 45 μg/g dry weight.

Nickel concentrations in the Nuduay River wetlands in the city of Monchegorsk, Russia, in 1982, 1985, and 1987 were [6, with additions]:

Water: 300 ± 24 μg/L (based on the 1982 OGSNK data, the annual mean concentration in river water in the city of Monchegorsk was 13200 μg/L [17])

Seston: 120 ± 15 μg/g dry weight

Bottom sediments: 540 ± 36 μg/g dry weight

Periphyton (*Cladophora glomerata*, *Oscillatoria* and *Cymbella ventricosa* species, sludge bacterial film): 300 ± 21 μg/g dry weight

Lumbricidae (without division into taxa): 220 ± 34 μg/g dry weight.

The frequency of reduced occurrence in periphyton was 50 to 80%. The total number of bacteria in bacterioplankton in the Nuduay River in the city of Monchegorsk in 1986-1992 was 6.4 x 10^6 cells/mL and the abundance of saprophytic microflora was 381 x 10^3 cells/mL [6].

5.5. THE KARA SEA HYDROGRAPHIC AREA

Wetlands of the Kara Sea hydrographic area are polluted with effluents of ferrous and non-ferrous metallurgy, electric power, chemical, oil refining, pulp mill, forest, and wood-working industries [6].

The Schuchya River wetlands (Lake Pyasino basin), Krasnoyarsk Krai, Russia, are probably the most heavily polluted wetlands in the territory of the former Soviet Union. The principal source of pollution is the Norilsk Integrated Iron-and-Steel Works. Copper is the principle metal pollutant. Copper concentrations in 1981 and 1987 were measured as follows [6, with additions]:

Water: based on the 1981 OGSNK data, annual mean copper concentrations in river water in the city of Norilsk reached 3310 μg/L [17]

Seston: 270 ± 32 μg/g dry weight

Bottom sediments: 900 ± 95 μg/g dry weight

Periphyton (*Melosira varians*, *Synedra ulna*, *Surirella ovata*, *S. angustata*, *Cymbella ventricosa*, sludge bacterial film): 800 ± 86 μg/g dry weight.

The frequency of reduced occurrence in periphyton was 80% or higher [6].

5.6. THE EAST SIBERIA HYDROGRAPHIC AREA

Water bodies of this hydrographic area are the most clean ones within the territory of Russia. The principal pollution sources are water transportation, municipal sewages, effluents of concentrating mills, and other enterprises. Due to low population density (less than 1 person per square km) the influence of these sources on water quality is insignificant [6].

5.7. THE CASPIAN SEA HYDROGRAPHIC AREA

The following major industries are located within the territory of the former Soviet Union: oil refining enterprises (the Volga-Ural, the North-Western, the Baku, and the North Caucasus complexes), metallurgic plants (the Southern and the Ural complexes), gas industry (in the Volga-Ural complex), non-ferrous works (the Ural and the Caucasus complexes), and chemical factories (the Volga, the Central, and the Ural complexes). Effluents of these industrial complexes, municipal sewages, and agricultural run-off are the principal sources of surface water pollution in the area.

The Chusovaya River wetlands (the Kama River basin) are downstream the city of Pervouralsk, Russia. The principal pollution sources at this site are the ash accumulation ponds of potassium bichromate works, copper smelter, and metallurgic plants [17]. Chromium, copper, and zinc are the principle metal pollutants. Chromium concentrations in the Chusovaya River wetlands in the period 1981 to 1985, and 1989 were [6, with additions]:

Water: 220 ± 32 μg/L

Seston: 540 ± 45 μg/g dry weight

Bottom sediment: 970 ± 81 μg/g dry weight

Periphyton (*Cladophora glomerata*, *Ulotrix zonata*, *Diatoma vulgare*, *Melosira granulata*, *Stephanodiscus hantzschii*, *Asterionella formosa*, *Microcystis aeruginosa*, bacterial film): 1200 ± 95 μg/g dry weight

Lumbricidae (without division into taxa): 320 ± 45 μg/g dry weight.

Copper concentrations in the same area were [6, with additions]:

Water: 43 ± 6 μg/L (based on the 1981-1985 OGSNK data, copper concentrations in the Chusovaya River water downstream the city of Pervouralsk varied from 67 to 388 μg/L [17])

Seston: 120 ± 12 μg/g dry weight

Bottom sediment: 750 ± 88 μg/g dry weight

Periphyton (*Cladophora glomerata*, *Ulotrix zonata*, *Diatoma vulgare*, *Melosira granulata*, *Stephanodiscus hantzschii*, *Asterionella formosa*, *Microcystis aeruginosa*, bacterial film): 150 ± 12 μg/g dry weight

Lumbricidae (without division into taxa): 230 ± 31 μg/g dry weight.

Zinc concentrations in the same area were [6, with additions]:

Water: 66 ± 8 μg/L (based on the 1981-1985 OGSNK data, annual mean concentrations in the Chusovaya River water downstream the city of Pervouralsk varied from 180 to 680 μg/L [17])
Seston: 200 ± 25 μg/g dry weight
Bottom sediment: 2400 ± 331 μg/g dry weight
Periphyton (*Cladophora glomerata*, *Ulotrix zonata*, *Diatoma vulgare*, *Melosira granulata*, *Stephanodiscus hantzschii*, *Asterionella formosa, Microcistis aeruginosa*, bacterial film): 220 ± 32 μg/g dry weight.
Lumbricidae (without division into taxa): 2200 ± 290 μg/g dry weight.

The frequency of reduced occurrence of periphyton was 60 to 75% [6].

5.8. THE PACIFIC HYDROGRAPHIC AREA

Wetlands of the Pacific hydrographic area are primarily polluted by the effluents of oil development, gas, mining, pulpmills, ship-building, and electric power industries as well as ferrous and non-ferrous metallurgy. Wetland quality in the southern areas of the Far East is impacted by agricultural run-off [6].

The Silinka River wetlands, Amur River basin, in the colony of Gorny, Russia, are an example. The principal pollution sources at this site are the ore mine "Solnechny" and effluents of the concentrating mill. Copper is the principal metal pollutant. Concentrations were measured as follows [6, with additions]:

Water: based on the 1981-1985 OGSNK data, annual mean concentrations in the Silinka River water downstream the colony of Gorny varied from 37 to 87 μg/L [17]
Seston: 85 ± 9 μg/g dry weight
Bottom sediments: 290 ± 35 μg/g dry weight
Periphyton (*Ulotrix zonata, Oscillatoria subtilissima, Synedra ulna, Melosira varians, Cymbella ventricosa, Nitzschia palea*): 110 ± 21 μg/g dry weight.
Lumbricidae (without division into taxa): 200 ± 28 μg/g dry weight.

6. Long-term changes of metal concentrations in soil organisms

Long-term dynamics of heavy metal concentrations in living organisms in the former Soviet Union could be studied using residue analysis of insect collections [18, 19, 20], mollusc collections [7, 8, 21], bone tissue cuts of roe (*Capreolus capreolus*) horns, moose (*Alces alces*) horns, and European deer (*Cervus elaphus*) horns [20], and plant collections [7, 22, 23].

6.1. PLANTS

Long-term dynamics of Pb and Hg in the Voronezh Biosphere Reserve wetland can be illustrated using data for two vascular plants (*Nuphar lutea* and *Menyanthes trifoliata*) which are abundant in Russian wetlands [7, 8, 23]. Herbarium and medicinal plants collected in 1913-1915, 1939, 1947, and 1958 in the wetlands upstream the Usman River and in 1977-1980 at the wetlands of the same river in the Voronezh Biosphere Reserve were used. The samples of *N. lutea* roots and *M. trifoliata* leaves were analyzed.

During the above mentioned period, mercury concentrations in the plants increased gradually from 0.03-0.04 to 0.12-0.15 μg/g dry weight. Lead concentrations in the plants increased as well but the rates of this increase were not uniform. From 1913-1915 to 1939, lead concentrations in the samples of both plant species did not increase (variations were not significant at the 0.05% level). From 1939 to 1947 there was a definite increase in lead concentration ($p < 0.001$). The increase in Pb concentrations in the plants might have been related to military actions in the region during World War II. This hypothesis was confirmed indirectly by the results of a study conducted from 1938 to 1952 in the wetlands of a

TABLE 1. Lead concentrations in Mollusca and Insecta collected in the Usman River Wetlands in the Voronezh Biosphere Reserve in 1930-1939 and in 1977-88 [1, 8]

Species	1930-1939 Concentration (μg/g)	1930-1939 No. Samples	1977-1988 Concentration (μg/g)	1977-1988 No. Samples
Mollusca				
Lymnaea stagnalis	1.8 - 0.1	6	6.4 - 0.4	12
Coleoptera (imagines)				
Discus marginalis	0.53 - 0.03	7	3.2 - 0.2	23
Carabus cancellatus	0.52 - 0.02	5	1.8 - 0.1	13
Carabus glabratus	0.62 - 0.03	5	1.8 - 0.1	13
Carabus marginalis	0.68 - 0.02	6	1.6 - 0.1	15
Anomala dubia	0.38 - 0.04	7	1.6 - 0.1	15
Acmaeops collaris	0.42 - 0.04	7	1.1 - 0.1	13
Silpha obscura	0.53 - 0.03	10	1.9 - 0.1	25
Silpha carinata	0.56 - 0.04	12	2.3 - 0.1	32
Heteroptera (imagines)				
Ilyocoris cimicoides	0.73 - 0.02	8	0.84 - 0.02	5
Notonecta glauca	5.4 - 0.3	51	3.7 - 0.2	46

region that was not influenced by military actions (Sysola River, city of Syktyvkar, Komi Republic, Russia). Lead concentration in the samples of this region definitely did not increase.

Variations in lead concentrations in plant samples within the period from 1947 to 1958 are uncertain, however, in the period from 1957 to 1977-1980 an increase of Pb was recorded again ($p = 0.02$)

6.2. ANIMALS

Analysis of lead concentration in mollusc and insect samples collected in the Usman River wetlands in the Voronezh Biosphere Reserve in 1930-1939 were 2 to 7 times lower than those of molluscs and insects of the same species collected at the Usman River wetlands in the Reserve in 1977-1980 (see Table 1).

The results presented in sections 6.1 and 6.2 clearly indicate that the ecosystem of the Voronezh Biosphere Reserve is exposed to anthropogenic metal pollution.

7. References

1. Zhulidov, A.V. and Emetz, V.M. (1996) Heavy metals, in *Former Soviet Union Region. Water Quality Assessment,* World Health Organization, Chapman & Hall, London, in press.
2. *State Report on Russian Federation Environment Status in 1993* (1994) The Russian Federation Ministry of Environmental Protection and Natural Resources, Zeleny Mir, Moscow, No. 24.
3. Nikanorov, A.M., Zhulidov, A.V, and Emetz, V.M. (1993) *Heavy Metals in Russian Wetland Organisms*, L. Hydrometizdat, St. Petersburg, 294 pp.
4. Zhulidov, A.V., Headley, J.V., Roberts, R.D., Nikanorov, A.M., and Ischenko, A.A. (1996) *Atlas of Russian Wetlands: Biogeography and Metal Concentrations*, NHRI Publishers, Canada, in press.
5. Cowardin, L.M., Carter, V., Golet, F.C., and Laroe, E.T. (1979) *Classification of Wetlands and Deepwater Habitats of the United States*, FWS30BS-79/31, Washington DC, 131 pp.
6. Zhulidov, A.V. (1995) Heavy metals as a factor of extreme impact on wetland ecosystems at the territory of the former Soviet Union, in *Life in Extreme Environments*, Proc. Toyama Int. Symp., 19-24 September 1995, Toyama University of International Study and Toyama Prefecture, Japan, pp. 39-61.
7. Nikanorov, A.M., Zhulidov, A.V., and Pokarzhevsky, A.D. (1985) *Biomonitoring of Heavy Metals in Freshwater Ecosystems*, L. Hydrometizdat, St. Petersburg, 144 pp.

8. Nikanorov, A.M. and Zhulidov, A.V. (1991) *Biomonitoring of Metals in Freshwater Ecosystems*, L. Hydrometizdat, St. Petersburg, 312 pp.
9. *Standard Methods for the Examination of Water and Wastewater* (1985) 16th Edition, Washington D.C.
10. Meybeck, D. Chapman, D., and Helmer, R. (Eds) (1989) *Global Freshwater Quality. A First Assessment,* Dlackwee Reference, 306 pp.
11. Sturgeon R. and Bergman S.S. (1987) Sampling and storage of natural water for trace metals, *CRC Critical Rev. Analyt. Chem.* **18**, 209-244.
12. Kowalsky, V.V. (1974) *Geochemical Ecology*, Nauka, Moscow, 300 pp.
13. Kowalsky V.V. (1978) Geochemical ecology - the basis of biogeochemical regionalization system, *Trans. Biogeochem. Lab.* **15**, 3-21.
14. Zhulidov, A.B. and Emetz, V.M. (1991) *Biogeochemical Regionalization of the USSR*, Manuscript, Hydrochemical Institute, Rostov-on-Don, 625 pp.
15. Zhulidov, A.V. (1990) Analysis of the factors determining the concentration of heavy metals in animal bodies as the bioindication of environment condition, in D.A. Krivolutsky (ed.), *Bioindication of Chemical and Radioactive Pollution,* Mir Publishers, Moscow, CRC Press, Boca Raton, pp. 251-303.
16. Lobchenko, E.E., Kolesnikova, T.X., Dolzhenko, A.S., Levina, M.E., Kuzhekova, N.I., Lyampert, N.A., Shaposhnik, E.G., Sorokina, E.F., and Pervysheva, O.A. (1988) *Dynamics of the USSR Surface Water Quality in 1976-80*, L. Hydrometizdat, St. Petersburg, 350 pp.
17. Lobchenko, E.E., Tsirkunov, V.V., Kuzhekova, N.I., Levina, M.E., Lyampert, N.A., Shaposhnik, E.G., Sorokina, E.F., and Pervysheva, O.A. (1991) *Dynamics of the USSR Surface Water Quality in 1981-85*, L. Hydrometizdat, St. Petersburg, 312 pp.
18. Zhulidov, A.V. and Emetz, V.M. (1979) Lead accumulation in a beetle body in case its habitat is exposed to pollution with automobile exhaust gases, *Trans. USSR Acad. Sci.* **244** (6), 1515-1516.
19. Nikanorov, A.M., Zhulidov, A.V., Emetz, V.M., Kulmatov, R.A., and Kist, A.A. (1982) Peculiarities of conjugated accumulation of mercury in water invertebrates and bottom sediments at the initial stage of metal accumulation in riverine ecosystems, *Trans. USSR Acad. Sci.* **264** (1), 1022-1024.
20. Zhulidov, A.V., Bogach, Y., Pokarzhevsky, A.D., and Gusev, A.A. (1985) Biomonitoring of the metal historic past of heavy

metals in natural ecosystems, in *Radioecology of Soil Animals,* Nauka, Moscow, pp. 186-198.

21. Nikanorov, A.M and Zhulidov, A.V. (1986) Background mercury concentrations in hydrobiota in the East-European subarctic lakes, in *Monitoring of Background Pollution of Natural Environment,* L. Hydrometizdat, St. Petersburg, 3rd. Issue, pp.186-190.
22. Zhulidov, A.V. and Scanlon, N.V. (1979) Variations in heavy metal concentrations in rare and endangered plant species in Yakutia and Buryatia in the 20-th Century, in *Ecological and Environmental Issues*, Kemerovo, Kemerovo University, pp.176-179.
23. Nikanorov, A.M. and Zhulidov, A.V. (1981) Biomonitoring of the historic past of river pollution with lead based on the study of metal accumulation in plants, *Trans. USSR Acad. Sci.* **258** (4), 1019-1021.

BIOINDICATOR SYSTEMS FOR SOIL POLLUTION: INSTRUMENTS FOR RISK ASSESSMENT

RECOMMENDATIONS
from a NATO Advanced Research Workshop held at the All-Russian Research Institute for Nature Protection, Moscow, 24-28 April 1995

The following recommendations are based on discussions held throughout the workshop and on specific questions raised after each presentation. A draft of the recommendations was distributed for comments to all participants before the final version was prepared. Therefore, the recommendations represent a consensus view among the various experts present.

Workshop participants recognized that the term bioindicator is used in variable senses, but did not discuss a rigid definition. There was a general agreement, however, on the fact that the term "bioindicator" refers to the ecological aspects, rather than to the physico-chemical aspects of an ecosystem, and that it has its main emphasis at the population and community level (presence or absence of species, diversity, trophic relationships, *etc.*).

Recommendation 1

Research should be stimulated to further develop bioindicator systems for the soil environment.

Although bioindicators have been used for a long time to evaluate the quality of the environment (specifically aquatic ecosystems), workshop participants agreed that their application to soils is still limited.

Recommendation 2

Soil bioindication should be based on information obtained from an array of different species or processes, jointly allowing an evaluation of the ecosystem as a whole.

As a new approach to developing soil bioindicators, workshop participants felt that these should not be restricted to one or a few species,

N. M. van Straalen and D. A. Krivolutsky (eds.), Bioindicator Systems for Soil Pollution, 249–253.

but should combine information from an array of different species or processes. This was emphasized in the word "system". A "system" of bioindicators is a collection of ecological information on an ecosystem, obtained in an organized way, such that the different components of information together allow conclusions on the quality of the ecosystem as a whole. This should include not only aspects of individual species, but also aspects of biodiversity and functioning. Given the great differences in sensitivity to pollutants among soil organisms, a broad diversity of species is of great importance.

Recommendation 3
An ecologically relevant bioindicator system should be based on information obtained from each of three categories: 1) invertebrate animals (macrofauna, mesofauna, nematodes), 2) microorganisms, 3) ecosystem processes (litter decomposition, rates of change in nutrient status, dynamics of soil food-webs).

The choice of suitable indicators is intimately linked to the function that the selected organisms have in the ecosystem. Workshop participants discussed the essential criteria for selecting bioindicator species and processes. In the selection process, a hierarchical approach may be followed, in which information on specific species of invertebrates is combined with information on microorganisms and soil processes. Microcosm studies were considered as useful experimental tools by which the impact on several organisms at the same time can be evaluated. The relative abundance of different feeding groups within a community was also considered as a promising endpoint for evaluation of effects.

Recommendation 4
Selection of bioindicators should allow for the different routes of exposure, in relation to the properties of chemicals and their metabolites, and the space-time scales of organisms in the soil ecosystem.

There are various ways in which organisms in soil may be exposed to potentially hazardous chemicals. The different exposure routes relate to the distribution of chemicals over the three main components of the soil ecosystem: solid phase, pore water, and air. These different exposure routes should be taken into account when composing a set of bioindicators. Exposure is also related to the spatial scale of the interaction between the organisms, the structure of the system, and the heterogeneous distribution of the chemicals. Bioindicators should not be restricted to the time-space scale of human dimensions.

Recommendation 5
A bioindicator system should integrate data on the concentrations of chemicals in organisms with data on the internal thresholds for ecotoxicological effects.

In the past, bioindicators have greatly relied on the chemical determination of concentrations of persistent chemicals in soil and in soil organisms. Workshop participants agreed that this is often a useful part of bioindicator research, but at the same time recognized that a bioindicator system should not be restricted to physico-chemical analysis. A possible way to combine data on residues of persistent chemicals in organisms with ecotoxicological effects is to calculate the ratio of the residue to the internal threshold concentration, determined in ecotoxicity and toxicokinetic experiments, and supported by field observations. In this way, problems of bioavailability differences between laboratory studies and the field may be overcome.

Recommendation 6
Bioindicators should be developed as doable tools, and should lend themselves easily for decision making processes.

The main aim of developing bioindicator systems is to support decisions on the management, clean-up or isolation of contaminated soils. Another field of application is the assessment of soil and sediment pollution as a factor limiting nature restoration projects. Bioindicators may be used for diagnosis of existing contamination, but also for prospective purposes. Because of this regulatory context, there should be a good link between the collection of scientific information and the use of that information in environmental policy. To facilitate this link, bioindicator techniques should be designed in an economical and practical way. Dissemination of information among scientists and decision makers can be improved by sharing information in world-wide computer networks.

Recommendation 7
Soil ecotoxicologists should exchange information on successful examples of the use of bioindicator systems, to find out the most effective ways to influence decision makers.

Workshop participants, being from 16 different nationalities, compared the various ways in which bioindicators are actually used to support environmental policy in different countries. The priorities given to certain environmental problems vary from one country to another (erosion, nuclear testing grounds, groundwater pollution by pesticides, quality of the urban environment). In many countries, the link between the scientific development of evaluation methods and environmental policy makers is not good or even absent. In some countries soil pollution problems are simply not addressed because no measurements are being taken. In other cases, information is collected, but is not followed up by relevant action. The potential for bioindicators to really influence environmental policy seems to be greatest in Germany, Denmark and the Netherlands. In the US, there is a tendency to use surrogate methods (obtained from aquatic toxicity testing) to arrive at decisions on contaminated soils. There was a general concern among the workshop participants about the present budget cut-downs in regulatory authorities charged with environmental protection.

Agreed, Amsterdam and Moscow, June 1995

Slava Artemov, ARRINP, Moscow, Russia
Jaroslav Bohác, Ceské Budejovice, Czech Republic
Marcel B. Bouché, INRA, Montpellier, France
Ruslan O. Butovsky, ARRINP, Moscow, Russia
Serge V. Chernyshenko, Dnepropetrovsk State University, Ukraine
Alexander Chetverikov, VNIIpriroda, Moscow, Russia
Vasily Demidov, ARRINP, Moscow, Russia
Jorge Domínguez, University of Vigo, Spain
Clive A. Edwards, Ohio State University, Columbus, USA
Alexei V. Essenin, ARRINP, Moscow, Russia
Jack H. Faber, IBN-DLO, Wageningen, Netherlands
Alexander Galagan, Institute of Geography, Kiev, Ukraine
Yuri Gorgo, Dnepropetrovsk State University, Ukraine
Hartmut Koehler, University of Bremen, Germany
V.M. Koropalov, ARRINP, Moscow, Russia
Gerard W. Korthals, Agricultural University, Wageningen, Netherlands
Dmitri A. Krivolutsky, Russian Academy of Sciences, Moscow, Russia
Paul Henning Krogh, NERI, Silkeborg, Denmark
Werner Kratz, Freie Universität, Berlin, Germany
Natalya Kuznetsova, Moscow State University, Russia
Roman G. Kuperman, Argonne National Laboratory, Argonne, USA
Ryszard Laskowski, Jagiellonian University, Kraków, Poland
Sergei Lomonosov, ARRINP, Moscow, Russia
Jean-Pierre Maelfait, Institute of Nature Conservation, Hasselt, Belgium
Janne Makaeva, ARRINP, Moscow, Russia
Pawel Migula, University of Silesia, Katowice, Poland
Lidiya Molodova, Gomel State University, Gomel, Belorussia
Elena Molozhanova, Norma Research, Kiev, Ukraine
Igor Netuzhilin, Russian Academy of Sciences, Moscow, Russia

Nina Osokina, Institute of Geological Sciences, Kiev, Ukraine
Maurizio G. Paoletti, Padova University, Italy
Rob W. Parmelee, Ohio State University, Columbus, USA
Andrei D. Pokarzhevskii, Russian Academy Sciences, Moscow, Russia
Almagul Rakhimbaeva, Kustanai Pedagogical Institute, Kazakhstan
Rotger Rippl, Vrije Universiteit, Amsterdam, Netherlands
David J. Spurgeon, University of Reading, United Kingdom
Georgios P. Stamou, University of Thessaloniki, Greece
Nico M. van Straalen, Vrije Universiteit, Amsterdam, Netherlands
Eugeny Tcherbakov, Russian Academy of Sciences, Moscow, Russia
Alexander Viktorov, Russian Academy of Sciences, Moscow, Russia
Vera Voznessenskaya, Russian Academy of Sciences, Moscow, Russia
Galina Zharova, Russian Academy of Sciences, Moscow, Russia

LIST OF CONTRIBUTORS

D.M. Bogomolov
The Ohio State University
Department of Entomology and
Environmental Science Program
1735 Neil Avenue
Columbus
Ohio 43210
United States of America

T. Bongers
Wageningen Agricultural University
Department of Nematology
Binnenhaven 10
NL-6709 PD Wageningen
The Netherlands

M.B. Bouché
Laboratory of Soil Animal Ecology
INRA/CNRS
BP 5051
F-34033 Montpellier, Cedex I
France

R.O. Butovsky
All-Russian Research Institute for
Nature Protection
Sadki-Znamenskoje
113628 Moscow
Russian Federation

S.K. Chen
The Ohio State University
Department of Entomology and
Environmental Science Program
1735 Neil Avenue
Columbus
Ohio 43210
United States of America

R.G.M. de Goede
Wageningen Agricultural University
Department of Terrestrial Ecology and
Nature Conservation
Kampsweg 27
NL-9418 PD Wijster
The Netherlands

J. Domínguez
University of Vigo
Department of Natural Resources and
Environment
Aptdo. 874
E-36200 Vigo
Spain

C.A. Edwards
The Ohio State University
Department of Entomology and
Environmental Science Program
1735 Neil Avenue
Columbus
Ohio 43210
United States of America

J.H. Faber
DLO/Institute for Forestry and Nature
Research
P.O. Box 23
NL-6700 AA Wageningen
The Netherlands

J.S.M. Heijmans
DLO/Institute for Forestry and Nature
Research
P.O. Box 23
NL-6700 AA Wageningen
The Netherlands

S.P. Hopkin
University of Reading
School of Animal and Microbial
Sciences
P.O. Box 228
Reading RG6 6AJ
United Kingdom

J.E. Kammenga
Wageningen Agricultural University
Department of Nematology
Binnenhaven 10
NL-6709 PD Wageningen
The Netherlands

H.H. Koehler
University of Bremen
Institute of Ecology and Evolutionary
Biology
D-28334 Bremen
Germany

G.W. Korthals
Wageningen Agricultural University
Department of Nematology
Binnenhaven 10
NL-6709 PD Wageningen
The Netherlands

W. Kratz
Free University of Berlin
Institute of Animal Physiology
Grunewaldstrasse 34
D-12165 Berlin
Germany

D.A. Krivolutsky
Russian Academy of Sciences
Institute of Evolutionary Animal
Morphology and Ecology
Leninsky Prospekt 33
117071 Moscow
Russian Federation

R.G. Kuperman
Argonne National Laboratory
Environmental Assessment Division
9700 South Cass Avenue
Argonne, IL 60439
United States of America

R. Laskowski
Jagellonian University
Department of Ecosystem Studies
Institute of Environmental Biology
Ingardena 6
30-060 Kraków
Poland

J.-P. Maelfait
Institute of Nature Conservation
Kiewitdreef 5
B-3500 Hasselt
Belgium

M. Maryanski
Jagellonian University
Department of Ecosystem Studies
Institute of Environmental Biology
Ingardena 6
30-060 Kraków
Poland

S. Mato
University of Vigo
Department of Natural Resources and
Environment
Aptdo. 874
E-36200 Vigo
Spain

P.J. Migula
University of Silesia
Department of Human and Animal
Physiology
ul. Bankowa 9
40-007 Katowice
Poland

M.G. Paoletti
Padova University
Biology Department
Via Trieste 75
35100 Padova
Italy

A.D. Pokarzhevskii
Russian Academy of Sciences
Institute of Evolutionary Animal
Morphology and Ecology
Leninsky Prospekt 33
117071 Moscow
Russian Federation

E. Pyza
Jagellonian University
Zoological Museum
Institute of Zoology
Ingardena 6
30-060 Kraków
Poland

R.D. Sandifer
University of Reading
School of Animal and Microbial
Sciences
P.O. Box 228
Reading RG6 6AJ
United Kingdom

D. Somaggio
Padova University
Biology Department
Via Trieste 75
35100 Padova
Italy

D.J. Spurgeon
University of Reading
School of Animal and Microbial
Sciences
P.O. Box 228
Reading RG6 6AJ
United Kingdom

G.B. Stamou
National Technical University of
Athens
Intelligent, Robotics and Control Unit
Zografou Campus
15773 Athens
Greece

G.P. Stamou
Aristotle University of Thessaloniki
Department of Ecology
PUB 119
540 06 Thessaloniki
Greece

S. Subler
The Ohio State University
Department of Entomology and
Environmental Science Program
1735 Neil Avenue
Columbus
Ohio 43210
United States of America

N.M. van Straalen
Vrije Universiteit
Department of Ecology and
Ecotoxicology
De Boelelaan 1087
NL-1081 HV Amsterdam
The Netherlands

J. Wojtusiak
Jagellonian University
Zoological Museum
Institute of Zoology
Ingardena 6
30-060 Kraków
Poland

A.V. Zhulidov
Hydrochemical Institute
Stachki Avenue 198
Rostov-on Don 344104
Russian Federation

SUBJECT INDEX